THE SKY AT NIGHT

Patrick Moore

THE SKY AT NIGHT

Patrick Moore

Foreword by Will Wyatt

W · W · NORTON & COMPANY
NEW YORK LONDON

Cover painting *Specially commissioned for a* The Sky At Night *episode, and never before reproduced elsewhere, this fine painting by Paul Doherty represents a singular black hole, perhaps towards the centre of the galaxy. The 'corona' effect is a result of the black hole drawing material towards itself from outer space, which energises into the visible spectrum as it approaches the event horizon.*

First American edition, 1986

Library of Congress Cataloging-in-Publication Data

Moore, Patrick
The sky at night.

Based on recent episodes of the television program The Sky at night.
1. Astronomy — Popular works. I. Sky at night (Television program) II. Title
QB44.2.M688 1985 520 86-12858

ISBN 0-393-30390-X

W. W. Norton & Company, Inc.,
500 Fifth Avenue, New York, New York 10110.

W. W. Norton & Company, Ltd.,
37 Great Russell Street, London WC1B 3NU.

1 2 3 4 5 6 7 8 9 0

CONTENTS

FOREWORD
by Will Wyatt

In April 1957 the price of a pint of beer was about 6d; Liverpool were in the second division of the Football League; and Colin Cowdrey and Freddie Trueman were preparing for a cricket season in which England beat the West Indies! The popular music of the day was dominated by skiffle, which featured the washboard and the string bass made from an old tea chest. A new kind of small car, the bubble car, was beginning to appear on the roads; the Mini was as yet unheard of.

The government of the day, presided over by Harold Macmillan, announced the end of compulsory National Service and introduced the continental idea of double white lines on the road to try to prevent dangerous overtaking. It was decided that work would definitely begin on the Channel Tunnel.

Further afield, the Suez Canal was reopened to shipping after the war of the previous year. Even further afield, the Americans attempted to launch the first earth satellite but the Vanguard rocket exploded on the ground (the successful Russian Sputnik was still six months in the future). And on April 24 1957, National Geophysical Year, Patrick Moore appeared on BBC 1 at 10:30 pm to present the first *The Sky At Night*. The programme has appeared every lunar month since, with Patrick Moore introducing every single edition. He must be the only television presenter to have had an asteroid named after him, Asteroid Moore.

Over those 27½ years he has communicated his enthusiasm for astronomy to countless viewers young and old, written over 60 books and become one of the best known faces in Britain. If astronomy has remained his chief claim to fame, it is not his only one. Over the past few weeks I have heard him reflect on his performances during the past cricket season, watched him take part in *Blankety Blank*, seen him play the piano accompaniment to the cellist Julian Lloyd Webber in order to raise money for the Children In Need appeal, read in a popular newspaper of his views on astrology ('bunkum') as well as heard and watched him explain the orange globe of Titan and the history of the Greenwich meridian on *The Sky At Night*.

I am indebted to the journalist Simon Hoggart for pointing out that if there is intelligent life in distant corners of the universe, the inhabitants may well have a particular interest in those monthly beamed signals from the planet Earth which are preceded by *At The Castle Gate* by Sibelius. In the star system Vega, which is 26 light years away, they will have been able to pick up the first 20 or so programmes. 'They must conclude', says Hoggart, 'that if Moore is a typical earthling (which he isn't), then we must be a terribly advanced life form.'

If the signals have failed to carry Patrick deep into space, he has, at least, been travelling busily in order to film special programmes for *The Sky At Night* viewers. His producer, Pieter Morpurgo, claims that Patrick must have one of the world's largest collections of suitcases, for wherever he goes his hosts press papers, souvenirs and other astronomical gifts upon him.

Patrick himself is unable to resist buying maps, plans, models and mementoes of the observatories he visits, and thus, invariably has to buy himself an extra suitcase in order to carry home his scientific loot. The chapters in this book reveal the extent of his travels over the past few years: to La Palma in the Canary Islands, home of the new British telescopes, to the Kennedy Space Centre in Florida, mission control at Houston, the United States National Observatory at Kitt Peak Arizona, down the great Homestake Gold Mine in South Dakota, up to the Mount Wilson and Mount Palomar observatories in California, to Mauna Kea, Hawaii, and to the Anglo-Australian Observatory in New South Wales and many more. Wherever Patrick goes he is welcomed as one of the leading exponents of astronomy. At each stop he has the same boundless enthusiasm to learn, to discuss and to pass on his knowledge to the viewers.

He is prone to floor the layman with his cheery cosmic view: 'We've been very fortunate, you know,' he is likely to say. 'There is an awful lot of debris floating about the Solar System and we haven't been struck by anything major for 65 million years.' I do not quite know how I should feel when he baldly tells me that 'the world *will* end when the sun swells into a huge red giant vaporising the inner planets'. But always with Patrick Moore there is a jolly conclusion. Before I have time to think, 'and me with them, I suppose,' he adds: 'but we've got five thousand million years to go, so we have no need to pack our bags yet'.

Now if Patrick, who has been timed speaking at 300 words per minute, were to talk to the camera non-stop for five thousand million years, that would be . . . No, I won't work it out. Let me thank Patrick for his great loyalty to the BBC and commend to you this latest collection of his work for the much cherished *The Sky At Night*. May his binocular lenses remain clear and unscratched for years to come.

Will Wyatt
Head of Documentary Features BBC TV

PREFACE

When I presented the first *The Sky at Night* television programme, in April 1957 (suitably ushered in by the Arend-Roland Comet), I had no serious thought that I would still be appearing monthly well over a quarter of a century later. Much has happened in the interval; after all, in 1957 the Space Age had not begun, and radio astronomy was at an early stage, while we did not even suspect the existence of exotic objects such as quasars, pulsars and scintars, to say nothing of Black Holes.

The past five years have been particularly eventful. There has been amazing progress in planetary studies; Saturn has been by-passed by the Voyagers, Venus has been extensively mapped by the US Pioneer and the Russian Veneras, while Voyager 2 is on its way to remote Uranus. IRAS, the Infra-Red Astronomical Satellite, has discovered thousands of new sources, to say nothing of possible planet-forming material associated with various stars, plus revelations much nearer home — dusty tails of comets, for example, and an extraordinary asteroid which ventures within nine million miles of the Sun. Quasars have come in for their full share of attention, and we are at last making progress in understanding them; pulsars continue to spring surprises on us, and a whole new class seems to have emerged. On the Earth's surface, the 'electronic revolution' has continued, and great new telescopes have come upon the scene; in particular the La Palma observatory has come into full operation. In space, the Shuttle has been successfully tested, and of course there has been the triumphal rescue of the ailing satellite Solar Max. Of special importance is the Starlink system, which will be of the utmost importance in the years to come. It seems strange now to reflect that not so very many decades ago, astronomy was widely regarded as a static science.

Compiling this book has not been easy. It is based, as before, on the television programmes, but although I have done my best to avoid overlap (which is why some months have been missed out) there is bound to be a certain amount of it, simply because I have tried to make each article self-contained.

I have been honoured by the presence of many distinguished astronomers on *The Sky at Night* programmes. My thanks are due also to Paul Doherty for his splendid illustrations, to those who have allowed me to use their photographs, to all at Patrick Stephens Limited (with special thanks to Bruce Quarrie and

Darryl Reach) and, as always, to Hilary Rubinstein.

I hope you will approve of the result. I wonder what will have happened by the time I start preparing the next volume of *The Sky at Night?*

Patrick Moore
Selsey, 1984

ACKNOWLEDGEMENTS

During the period covered by this book (April 1980 to August 1984) many guests have appeared with me on *The Sky at Night*. I am most grateful to them. In chronological order, they are: Dr P. Williams; Dr John McGraw; Heather Couper; Dr Keith Pierce; Dr Jack Harvey; Professor Geoffrey Burbidge; Professor Alec Boksenberg; Dr Garry Hunt; Dr Paul Murdin; Dr Bradford Smith; Dr Steve Peyander; Dr Torrance Johnson; Dr Edward Stone; Dr Hal Masursky; Dr Keith Matthews; Dr Eric Persson; Professor Gerry Neugebauer; Dr John Beckman; Caroline Herschel; Professor F. Graham Smith; Dr David Hughes; Professor Sir Bernard Lovell; Philip Henley; Dr David Allen; Dr David Malin; Dr Victor Clube; Commander Eugene Cernan; Dr Jim Emerson; Dr John Jeffries; Alan Pickup; Dr Terry Lee; Dr Dale Cruikshank; Dr René Racine; Dr Raymond Davies; Dr Keith Rowley; Iain Nicolson; Dr Hartmut Aumann; Dr Fred Gillett; Dr Robert Howard; Dr Douglas Duncan; Dr Olin Wilson; Dr Michael Penston; Dr Robert Fosbury; Dr Jasper Wall; Dr Peter Cattermole; Dr Simon Mitton; and Dr John Davies. To all these — thank you indeed.

1 THE MULTIPLE-MIRROR TELESCOPE

High in the Santa Rita Mountains of Arizona, between eight and nine thousand feet above sea level, you will find the Whipple Observatory, which boasts of the world's most revolutionary telescope. It sits on the very summit of Mount Hopkins; from below, the observatory seems to be precariously placed, and certainly the mountain road which leads up to it is not one to be negotiated by a nervous driver. The surface is reasonably good, but there are no safety-rails, and the road is narrow in places.

At first sight the observatory looks like nothing more than a huge, almost square hut. It bears no resemblance to the graceful domes of, say, Palomar or Mount Wilson, and it is indeed built upon a completely different principle. As the telescope moves round, so the whole observatory turns with it. Even though the building weighs 500 tons, the motion is amazingly smooth.

The instrument is known as the MMT — the Multiple-Mirror Telescope. Instead of one immense mirror to collect light, there are six. Each is of 'egg-crate' construction, and is 72 in in diameter; each weighs only about one-third as much as a conventional mirror of the same size. They work together, so that the light they collect is brought to the same focus, and is combined. The overall result is equal to a reflector with a 176 in mirror, and there are only two telescopes in the world larger than that: the Palomar 200 in, which was made over 30 years ago now, and the Russian 236 in, which is frankly unsatisfactory and has not yet produced any worthwhile results.

The MMT does not even look like a telescope. It is short and squat, with a maze of girders and wires, in the midst of which are the six 72 in primaries. There is also a separate 30 in mirror which is not connected with the main system, and which acts as a guide. All the mirrors are silica-plated, and are accurate even by the exacting standards of today.

The idea of using several mirrors in unison to produce a single image is not new, but until recently it was quite impracticable, because of the immense difficulties of guiding and lining-up. Without the very recent developments in computer techniques, the MMT could not have been made. There is also another all-important feature of the design: the active laser control system.

Keeping even a few mirrors in true alignment, as with a conventional reflector, is difficult enough, and the problems are enormously magnified with the MMT. If any one of the six primaries is fractionally out of position, then the telescope will not perform properly. Manual adjustment is out of the question, and so a laser is used. A laser may be described as a pencil-thin beam of light, known technically as coherent light, which does not spread out, and can therefore be very accurately positioned. The transmitter is used to send a laser beam through the optical system, ending up by producing what may be termed an artificial star. The required position of this artificial star is known. If the image is formed exactly where it ought to be, then everything is lined up, and the focus will be correct. If not, an automatic adjustment is made. The laser control has to be kept in operation all the time that observations are being made, because conditions alter whenever the telescope shifts; gravitational strains, for instance, are quite enough to produce unacceptable distortions unless immediate compensation is made.

The mounting is of the altazimuth type. That is to say, the telescope can move either up and down (in altitude) or round (in azimuth), and both these movements have to be taken into account when the telescope is following an object in the sky. Previous large telescopes have usually been mounted equatorially, so that the altitude movement looks after itself, but this would be impossible with the MMT, so that modern computer techniques are needed. Actually, the Russian 236 in also has an altazimuth mounting — its one major contribution to date.

At an early stage in the planning of the MMT, the decision was taken to make the entire observatory rotatable. This means that the slit is always in front of the telescope. The two are driven independently, but there is an automatic safety device to prevent any possibility of a collision. The whole observatory is so skilfully built that a small motor, developing less than 20 horse-power, is adequate to rotate it. Obviously the tracks for the wheels have to be kept clear, and snow can be a problem. Recently there was another unexpected hazard — a plague of ladybirds, which were everywhere. Astronomers found to their dismay that the smooth rotating of the observatory was being upset by the myriads of

My photograph of the MMT as we were making our programme.

squashed ladybirds in the tracks, and special sweepers had to be introduced to cope with the situation. They worked well, and it may be added that they also proved able to deal with extra hazards such as dropped screwdrivers . . .

The MMT was not built as a single unit. For instance the main mounting was made in Italy, the steel parts in Canada and the optics in California. Transporting them to the summit of Mount Hopkins, up that alarming road, was in itself quite a major problem, and on the whole it is surprising that the telescope was brought into action on schedule. There is constant maintenance; all the procedures are completely different from those with an ordinary telescope.

This being so, why build so complex an instrument? There are several reasons, one of which is sheer cost. The MMT is much cheaper than a single-mirror 176 in reflector would have been. Moreover, there are positive advantages. The MMT is a prototype — the first of a whole new generation of telescopes. It may well be that mirrors of much over 200 in aperture will present insuperable problems, however well they are mounted (the Russian colossus is a case in point), and by using smaller mirrors in unison this kind of hazard is greatly reduced. If it were possible to build an MMT with primaries of (say) 100 in aperture, the result would be an instrument with a greater light-grasp than anything so far built, and there is no reason why this should not be done. Indeed, plans are already being drawn up.

Astronomers who have used the MMT are unanimous in praising it. The end product, so far as they are concerned, is much the same as with a normal telescope, and some very interesting results have already been obtained. For example, Dr John McGraw has been carrying out careful studies of a very peculiar star, known by its catalogue number of PG 1159-035, which is only of the fifteenth magnitude and is therefore much too faint to be examined with any but a giant telescope. The star is pulsating, with periods of five and eight minutes. This is the kind of behaviour which would normally be associated with a very small, super-dense star of the White Dwarf variety. But PG 1159-035 is not a White Dwarf — its surface temperature is extremely high, at least 50,000°C, possibly as much as 70,000°C, as against the modest 6,000°C of our own Sun. Moreover, it is very distant — approximately 700 light years away from us. At the moment its exact nature is uncertain, but there is every possibility that it is evolving rapidly, and may show permanent changes over a period of only a few years. This, of course, is something entirely new. Most stars change much too slowly for their life-stories to be followed directly; our Sun has not altered much for several thousands of millions of years, and will not do so for several thousands of millions of years in the future. With PG 1159-035 we may be watching a star as it evolves in front of our eyes, and if this is confirmed it will be of tremendous importance in theoretical astronomy.

The MMT has been operating for only a few years, but it has already proved its worth. Its designers have said that if they had to start all over again, there is surprisingly little that they would alter. Others of its kind will soon follow, and in the history of observational astronomy the Mount Hopkins telescope will probably rank as being as important as Galileo's first refractor, Newton's

The top of the MMT.

original reflector, or the great telescopes of Mount Wilson and Palomar. Utterly unconventional, strange in appearance and housed in its remarkable rotating 'box' on the top of the peak, it has shown that the real future of Earth-based astronomy may lie with the Multiple-Mirror Telescope.

2 JOURNEY TO THE CENTRE OF THE GALAXY

Look low down in the southern sky, on a summer's evening, and you will see the magnificent star-clouds in the constellation of Sagittarius, the Archer. Even with the naked eye they are conspicuous; binoculars show that they are made up of stars, so numerous that to count them would be a hopeless task. The stars look so crowded that they seem to be in imminent danger of collision. Yet as so often happens in nature, appearances are deceptive; the stars are widely spread-out, and we are dealing with nothing more significant than a line-of-sight effect.

Beyond the star-clouds of Sagittarius lies the centre of our Galaxy, around 30,000 light-years away from us. The Galaxy is a flattened system, with a central bulge or nucleus; I have compared it, graphically if unromantically, with the shape of two fried eggs clapped together back to back. When we look along the main plane we see the effect of the Milky Way, with many stars almost one behind the other. Not surprisingly, the Milky Way is particularly rich in Sagittarius.

What lies at the very centre of the Galaxy? This is what astronomers would very much like to know, and recent results have been of immense interest; but one way we cannot hope to find out is by sheer visual observation. Light-rays cannot pass through the dust which lies between the galactic centre and ourselves — in fact the brightness of an object near the centre would be reduced by 28 magnitudes, far too faint to be detected with any telescope. We must therefore rely upon 'invisible astronomy', in particular radio astronomy and infra-red astronomy.

For many years it has been known that the Galaxy is spiral in form, but this is not surprising, because many other spiral galaxies are known. A good example is Messier 31 in Andromeda, which is a mere 2.2 million light-years from us, and is a member of our Local Group of systems. Messier 31 is dimly visible with the naked eye; photographs taken with large telescopes show its well-marked spiral form, rather like a Catherine-wheel, though unfortunately it lies at a sharp angle to us, so that its full beauty is lost. Messier 51, the 'Whirlpool' in the constellation of Hunting Dogs, is face-on, and the spiral shape is beautifully displayed. We also have barred spirals, in which the arms seem to come from the ends of a 'bar' through the nucleus, as well as elliptical and even irregular systems. Some galaxies are very active, with indications of great disturbances in or near their centres; this is true of some of the ellipticals, as well as the spirals with bright, condensed nuclei and weak arms, known as Seyferts in honour of Carl Seyfert, who drew attention to them as long ago as 1942.

There have been suggestions that some very violent object lies at the centre of our own Galaxy, and certainly there is a strong radio source there. In general the arms of our spiral rotate round the nucleus at uniform velocity, but in 1957 the Dutch radio astronomer Jan Oort found that one arm showed a definite velocity of approach toward us at a rate of 50 km per second. It is known as the '3 kiloparsec arm'; a kiloparsec is equal to 1,000 parsecs, and 1 parsec is equal to 3¼ light-years, so that the approaching arm is some 9,000 light-years from the galactic centre.

Oort went on to show that there are other arms and clouds, closer to the centre, which are similarly expanding outward at velocities ranging up to 100 km per second, which is nearly a quarter of a million miles per hour. There is also an immense disk of cool hydrogen which is non-expanding, together with a ring of clouds containing molecules.

Molecules, or atom-groups, are comparatively fragile, and can survive in these relatively dense cool clouds only when undisturbed. Many varieties have now been identified — at least 60 in all, including formaldehyde, carbon monoxide and even alcohol. The ring of clouds containing these molecules is expanding outward at 160 km per second. Mixed up in the clouds there are also what are termed HII regions, made up of hydrogen atoms which have been broken up by powerful radiation. Many HII regions much closer to us are known; the Orion Nebula is one, and they are known to be regions in which fresh stars are being produced from the interstellar material. It seems, then, that star-formation is going on in areas surprisingly close to the galactic centre.

Inside the ring there is a cloud of hydrogen at a higher temperature, known as the Arc because of its shape; this is moving outward at 50 km per second. And still nearer the centre there is the remarkable and significant object known as Sagittarius A. It consists of two components, an eastern and a western. Between them (on the far side of the galactic centre, as seen from Earth) there is a cloud known as the '40 km/sec' feature; it is often believed to have been ejected from

The Milky Way — a composite picture from the Lund Observatory in Sweden.

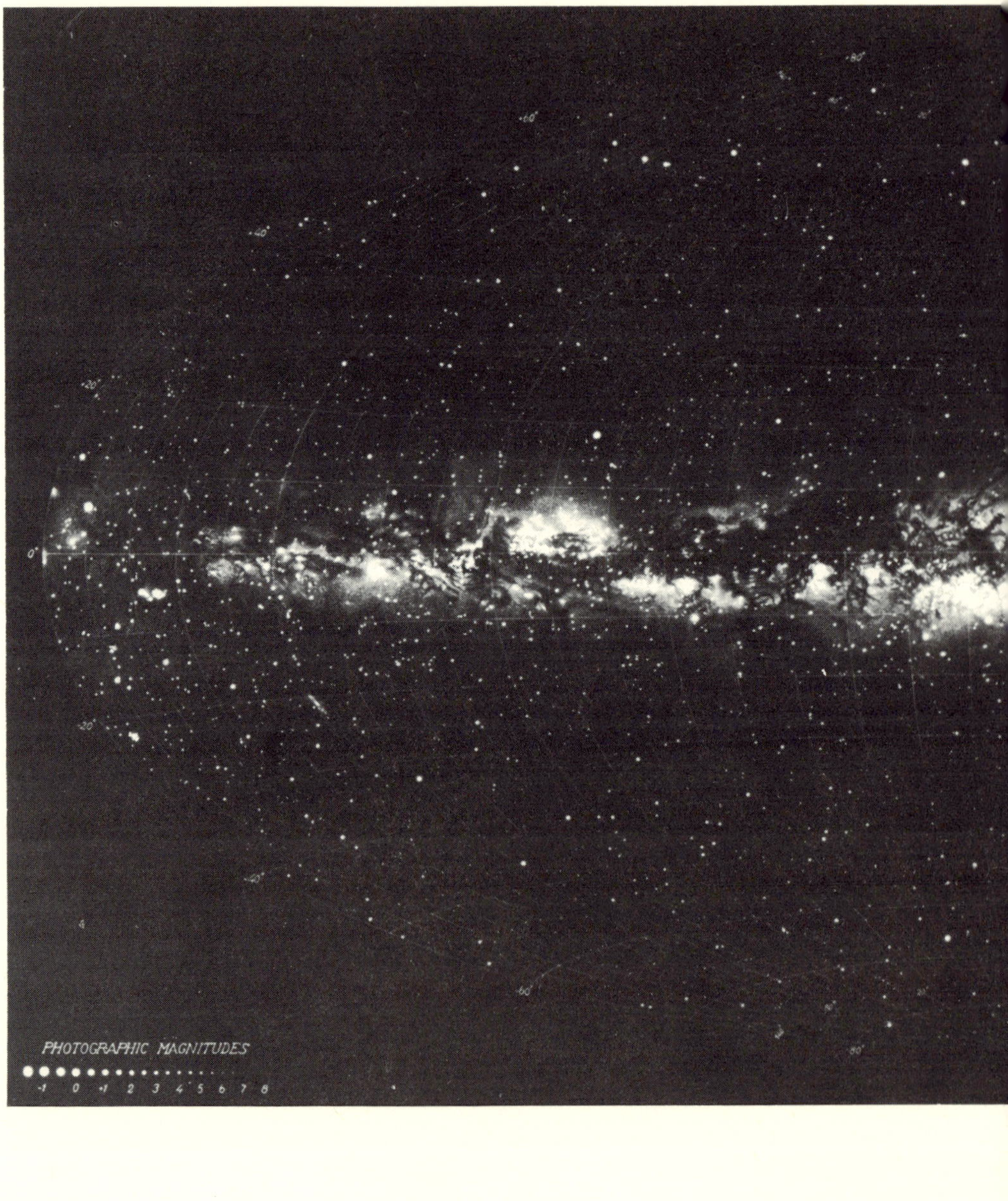

the centre of the Galaxy only about half a million years ago, so that star-formation inside it has not had time to begin. After all, half a million years is a very short period by cosmical standards.

Sagittarius A East is an ordinary supernova remnant — all that is left of a massive star that exploded in the remote past. (Supernovæ have often been observed in other galaxies, and occasionally in our own — the last to be seen in

The globular cluster Messier 5, in Serpens. It is about 30,000 light-years away and 100 light-years in diameter. It is easily seen in binoculars.

our Galaxy was Kepler's Star of 1604.) Sagittarius A West, on the other hand, is an evenly-lit HII, region which contains yet another disk possibly illuminated by a single central source. This disk is approximately four light-years in diameter, which is about the same as the distance between the Sun and the nearest stars, those of the southern Alpha Centauri system. It is not expanding, it is rotating uniformly, and is rich in dust and in young stars.

Though the whole region is quite invisible optically, it has been possible to use infra-red techniques to measure the star-density close to the galactic centre. The stars are concentrated into a dense, central cluster. There are about a million of them altogether, mainly elderly and orange or orange-red, packed a hundred million times more closely than the stars in our part of the Galaxy. They are separated not by light-years, but by only a few light-days. If we lived on a planet circling a star in this cluster, the night sky would be superb indeed; many of the stars would show up as disks, and there would be no darkness at all.

The total number of the stars in this region can be estimated from the amount of radiation emitted, but one problem arises straight away. The combined mass of these stars is not enough to prevent the disk from dispersing. There should be about five times as much. In fact, we have come up against the question of the 'missing mass'. It must be present — but where? Simple non-luminous clouds of material can hardly be the answer, and it is more likely that the mass is contained in the central radio source in the heart of the cluster. This source has been found to be about three times larger than the diameter of Neptune's orbit round the Sun, but a high proportion of its power comes from an ultra-compact region less than one thousandth of a second of arc across — roughly the size of the orbit of Jupiter, between 900 and 1,000 million miles. Within this relatively tiny area we find an amount of energy comparable with that coming from the central regions of active galaxies many millions of light-years away. Remember, too, that we are dealing with an upper limit; the true diameter may be considerably less than that of Jupiter's orbit.

If this is correct, then we have at last come to the centre of the Galaxy, and to the source which is illuminating Sagittarius A West. And although we must now abandon observation, and turn to theory, it is conceivable that the source is nothing more nor less than a Black Hole.

Black Holes have been very much in the news recently. If a very massive star collapses abruptly, and becomes extremely small and super-dense, it may pull so strongly that not even light can escape from it, so that it is surrounded by a kind of 'forbidden zone' which is to all intents and purposes cut off from the outside universe. It is suggested that Black Holes may form near the cores of galaxies — and the more massive the galaxy, the bigger the Black Hole. Some vast galaxies, such as the strange, active Messier 87 in Virgo, may have central Black Holes with 1,000 million times the mass of the Sun, whereas our own more modest Galaxy can manage a Black Hole with no more than five million times the Sun's mass. But if it exists, it could provide enough mass to make the ionized disk of hydrogen stay stable, and it would certainly give a logical solution to the problem.

In this case, we are entitled to speculate further. The Black Hole acts rather in the manner of a cosmic plughole; matter will swirl outside it in its gravitational field before being swallowed up and this swirling matter will form an 'accretion disk' which is hot and glowing, emitting radio waves and higher-energy radiations. Because it is spinning rapidly it generates shock-waves round it sufficient to drive streams of gas outward.

Undoubtedly there is something in the galactic centre which is responsible for the expanding arms and clouds, but it seems that the accretion disk itself is not active at the present moment, because it does not contain enough mass. Yet suppose that it showed more activity in the past? It could have expelled the 3-kiloparsec arm about 13 million years ago, and the 40 km per second feature half a million years ago. In the future, it could become active again if enough material could collect in the accretion disk.

All this is highly controversial, and there are some authorities who hold very different views, but it is at least a possibility.

We need have no fear that our Solar System will be swallowed up by a Black Hole, because we lie at a reassuringly safe distance. We cannot even be sure that a massive Black Hole exists at the centre of the Galaxy. But at least there is little doubt that something very strange lies hidden in those mysterious, far-away regions in the very heart of our star-system.

Overall view of Kitt Peak, showing the domes for the optical telescopes.

3 THE SUN – FROM KITT PEAK

To us, the Sun is the most important body in the universe. It sends us virtually all our light and heat, and without it we could not exist for a moment; indeed, the Earth itself could never have come into being. Yet even though it is so close to us on the astronomical scale — a mere 93 million miles on average — there is still a great deal about it that we do not know. Studies of the Sun are not nearly so straightforward as used to be thought a few decades ago.

Solar research involves using equipment of a specialized kind. Much the largest solar telescope in the world has been set up at Kitt Peak, near Tucson in Arizona, and it is from here that many results of outstanding importance are being obtained.

The story of Kitt Peak really began as long ago as 1950, when the National Science Foundation was considering the setting-up of a new major astronomical research centre. High altitude above sea-level is usually an advantage (the astronomer is always anxious to climb as high as he can above the dirty and unsteady lower layers of atmosphere), and over 50 sites were examined. Kitt Peak seemed to have many points in its favour. Its height is of the order of 7,000 ft; weather conditions are usually good, and Tucson is not far off, though the distance is great enough for the city lights to be unobtrusive. There were, of course, other points to be settled. Kitt Peak comes into the reserve of the Papago Indians, and the sacred mountain Babuquivari is regarded by them as the centre of the universe, with the gods living in convenient nearby caves. After prolonged negotiations, everything was settled; the Observatory acquired the lease, with the guarantee that the sacred caves would never be disturbed — a promise that will most certainly be kept.

The first dome on Kitt Peak was erected in 1958, and others have followed. Today there are 17 in all, containing powerful optical telescopes; reflectors of 24 in, 36 in, 50 in, 84 in and 90 in aperture, and finally the great 158 in (4 m) telescope, which is one of the largest and best in the world. The solar telescope is quite different, and at first sight it does not look like a telescope at all; it gives the impression of being a huge, white, inclined tunnel. Its whole design is revolutionary, but it has well and truly its worth.

The upper mirror, 80 in in diameter, is called the heliostat. It can be rotated; it catches the Sun's light, and directs the rays down the 600 ft tunnel in a fixed direction. At the bottom of the tunnel is another mirror, 60 in in diameter, which is curved, and which reflects the rays back up the tunnel to the 'half-way house', where there is yet another mirror — flat, this time — which sends the rays down through a hole into the solar laboratory, where the analyses are carried out. The great advantage of this arrangement is that the sunlight always arrives at the same point, so that the heavy analytical apparatus does not have to move at all; indeed, the only movable mirror in the whole telescope is the upper heliostat. Something of the same kind can be used with large optical telescopes, and is known as the Coudé system; for instance, the Kitt Peak 158 in telescope has a Coudé focus which is frequently used.

Few solar telescopes are anything like as large as this, because with the Sun there is plenty of light available; the Kitt Peak instrument is much bigger than any other solar telescope in the world, and there are various advantages. The Sun's image in the laboratory is almost 3 ft in diameter, and there is so much light that it can be spread out into a very long spectrum. With other 'tower telescopes', with rotating heliostats, the tunnels are usually vertical, but with the 600 ft Kitt Peak tunnel this would involve an alarmingly great height, which is why it has been set at an angle (even so the heliostat is 100 ft above the ground, and from the upper platform the view of the surrounding countryside is magnificent). There is an escalator inside the tunnel, and riding up and down is easier than negotiating the dozens of steps! Temperature control is an important factor. If there were marked fluctuations, the air inside the tunnel would be unsteady, and this would ruin the telescope's performance, so 1,700 gallons of cooling fluid are circulated between the inner and outer walls of the tunnel, keeping the temperature inside constant to within a degree or so.

Originally it had been planned to lower the heliostat into the top of the tunnel during periods of bad weather. The cost was estimated at 75,000 dollars, but when this rose to a quarter of a million dollars the scheme was abandoned, and the heliostat is simply covered up when necessary. It seems to come to no harm, even when there is heavy snowfall (as happened when I went to the Peak on one occasion in 1980).

The Sun is a star. The temperature of the bright surface or photosphere is almost 6,000°C while near the core the temperature rises to the incredible value of about 14 million °C or perhaps even more; the solar energy is produced by nuclear reactions — basically, hydrogen is being converted into helium, with the release of energy and loss of mass. The Sun loses about four million tons per second, though fortunately there should be no major change in output for at least 5,000 million years in the future. By stellar standards the Sun is still no more than middle-aged.

Darker patches, known as sunspots, are often seen on the surface, and are centres of powerful magnetic fields, sometimes 3,000 to 5,000 times as strong

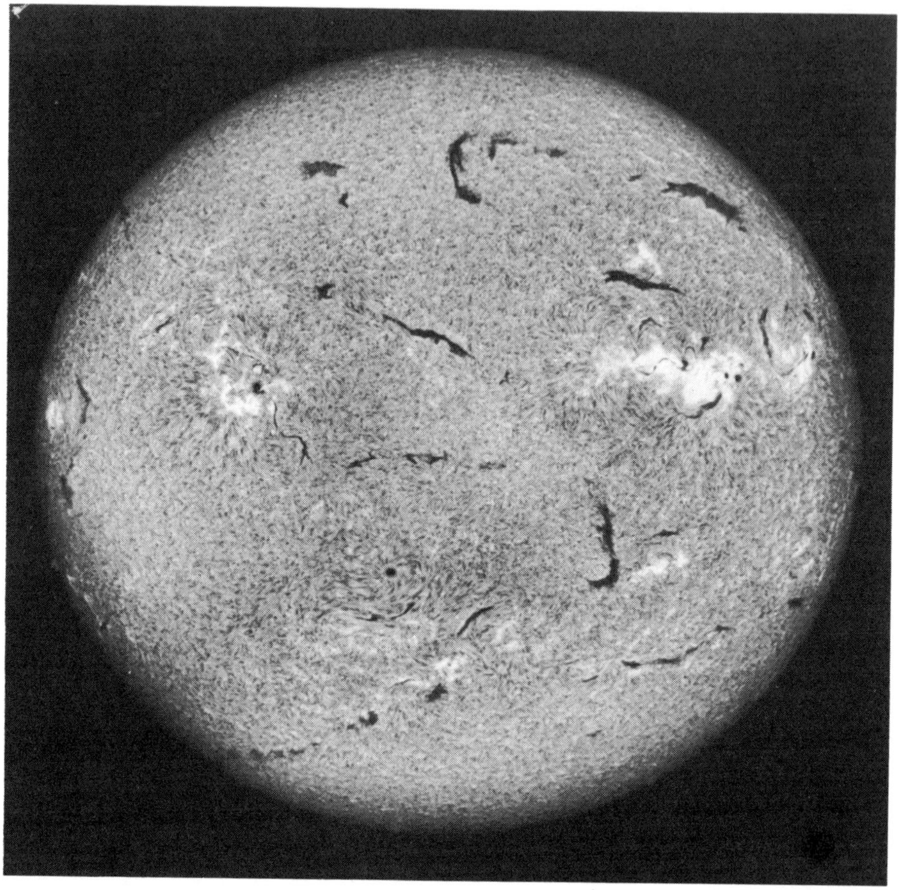

Photograph of the Sun taken in the H-Alpha Line with the Razdow Solar Optical Telescope, WS250–1A Solar Observatory, Ramey Air Force Base, Puerto Rico, July 15 1968.

as the magnetic field of the Earth. Yet rather surprisingly, the Sun's overall magnetic field is negligible, and is merely made up of the fragments of the magnetic fields of sunspot groups that have disappeared (no spot lasts for more than a few months, and many exist for only a few days or even less). Investigations of the magnetic properties of the Sun form an important part of the solar programme at Kitt Peak, and here the great size and light-grasp of the telescope is a major factor. Unquestionably the spots are magnetic phenomena, though it cannot honestly be said that we really understand how they originate.

The Sun may be regarded as a variable star, because there is a well-marked cycle of activity. Maxima, with numerous spots and associated phenomena, occur at intervals of about 11 years, though the period is by no means constant;

there was a maximum in 1957–8 and another in 1969, with another in 1980. However, we cannot be sure that the cycle is a permanent characteristic of the Sun, and there are strong indications that between 1645 and 1715 spots were very rare, in which case the cycle was to all intents and purposes suspended.

There are other puzzling points, too. Theoretical considerations indicate that the Sun should send out large numbers of neutrinos, which are particles with no electric charge and virtually no mass, so that they are very difficult to trap. The best method is to try to 'catch' them in a large underground tank of what is simply cleaning fluid! On the rare occasions when a neutrino interacts with one of the chlorine atoms in the tank (and cleaning fluid contains a great deal of chlorine), it produces effects which can be measured. What worries astronomers is that the numbers of neutrinos seem to be very much less than expected, so that something is wrong somewhere. It has even been suggested that the Sun's energy has been 'switched off' so to speak, for the moment, though it is more likely that we lack sufficient knowledge of the ways in which neutrinos behave. Subterranean 'observatories' of this kind have been established in the United States (at Homestake Mine, South Dakota) and in the Soviet Union. Research is being carried on energetically, and the work at Kitt Peak is closely bound up with it.

Though the telescope was built for studies of the Sun, it can also be used for observing the stars, particularly for research into their magnetic fields. Of course, it is not so large as some other optical telescopes — the main mirror is 'only' 60 in in diameter — but once again, the fact that the light is brought to focus at a fixed point is a great advantage, because the heavy apparatus used for analysis does not have to be moved around. Consequently the telescope is in use both day and night, and is probably the busiest instrument upon Kitt Peak.

Today Kitt Peak is regarded as the national observatory of the United States,

Sunspots, photographed by the late W. M. Baxter with his 4 in refractor.

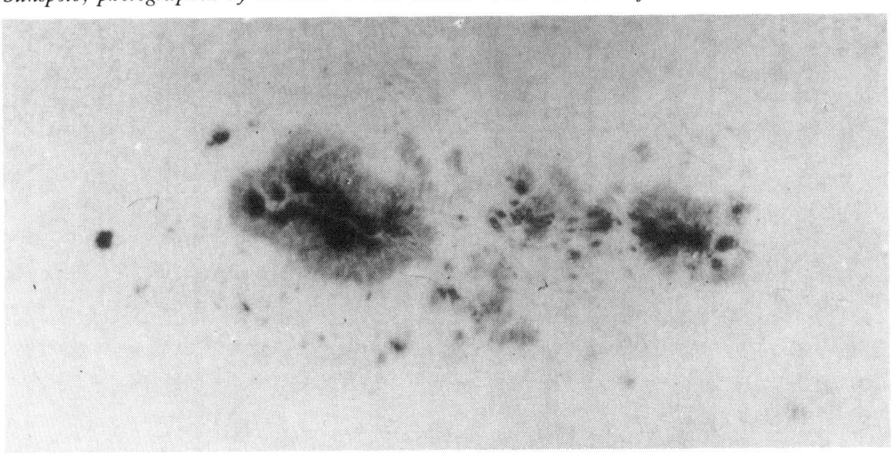

and Tucson is an astronomical centre second to none. As has been pointed out by the British-born Director, Dr Geoffrey Burbidge, visiting astronomers from overseas countries are given about half the observing time with the major telescopes. In the future it is hoped to add great new instruments. In any case, Kitt Peak is certain to remain in the forefront of scientific research — even though the Papago Indians were mistaken in their belief that the sacred mountain close by is the centre of the entire universe!

4 METEOR CRATER

Go to the desert of Arizona, not so very far from the town of Flagstaff, and you will find perhaps the most majestic natural monument in the world. It is a huge crater, over 4,000 ft in diameter, with sides sloping down steeply to its sunken, bowl-shaped floor. We know it as Meteor Crater; more properly it should be called Meteor*ite* Crater, because it was formed by a tremendous iron missile which hit the desert approximately 22,000 years ago. Svante Arrhenius, a great Swedish scientist of the early 20th century, aptly described it as 'the most interesting spot on earth'.

The first white men to find the crater, in 1871, rather naturally regarded it as volcanic in origin; there are volcanoes in the same area — notably Sunset Crater, which erupted as recently as 1250 AD, and is still classed as active. Twenty years after the discovery, G. K. Gilbert, one of America's leading geologists, visited the site and described it as 'a steam explosion of volcanic origin'. Then, in 1903, the crater was carefully explored by Daniel Moreau Barringer, a Philadelphia mining engineer. Barringer did not agree with Gilbert's theory; he became convinced that a meteorite had been responsible for blasting out the crater. In this case it was reasonable to assume that the meteorite itself must be buried somewhere underneath. Iron is valuable; Barringer acquired the land by purchase, formed a company to exploit it, and set to work to locate the hidden missile.

Because the crater is so nearly circular, Barringer believed that the missile must have come straight down, in which case it would presumably be found near the centre of the bowl. Mining operations were started in 1906, and went on for a year, but the results were not encouraging. Small fragments of meteoritic material were found, but the underlying water-table had combined with shattered sandstone to form a mixture which made mining very difficult indeed.

Barringer made new calculations, and decided — correctly, this time — that a large impacting missile would produce a circular crater even if it came in at an angle. He concluded that the meteorite had approached from the north, at an angle of around 87°, and had buried itself below the south-eastern rim of the crater. Drilling was re-started, and numerous meteoritic fragments were unearthed, but progress was slow, sometimes no more than 18 in to 20 in in a

Meteor Crater as I photographed it from the air.

full day's work. Finally, at a depth of less than 2,000 ft, the drill jammed. Nothing more could be done; the main effort came to an end in 1921. Spasmodic efforts were continued until Barringer's death in 1929, but by then it had become clear that the meteorite was out of reach.

Since then, the only attempts at commercial exploitation have been in connection with silica mining, since the silica here — particularly in the south — is as pure as any in the world. Inevitably there was damage to the crater, and, not to put too fine a point on it, the miners were an unmitigated nuisance. In 1967 the site was adopted as a National Natural Landmark, and concerted action prevented any resumption of the silica mining, as was threatened in 1970. To spoil the crater would indeed be a crime; there is no other impact structure anything like so perfect or so spectacular.

Modern researches have thrown considerable light upon the nature not only of the crater, but of the meteorite which produced it. The rim-to-rim diameter is 4,150 ft, giving a circumference of three miles, the depth is 570 ft. From the outer plain the view is not impressive since the rim rises to only a very modest altitude — which makes the view even more impressive when one suddenly comes upon it. A small but excellent museum has been set up on the rim, containing explanatory diagrams as well as specimens of meteoritic fragments found in the area; visitors can walk along the rim for some distance, and it is even possible to follow the so-called 'trail' right down to the floor, though special permission has to be obtained because of the steepness of the descent. It

Above *The walls of Meteor Crater.*

Below *The floor of Meteor Crater.*

takes less than an hour to climb down, but considerably longer to return, particularly under the broiling heat of the Arizonan sun!

According to well-informed estimates, the meteorite was around 90 ft in diameter, though the solid mass which finally impacted had been reduced to 80 ft or so. It struck the ground at a velocity of 30,000 to 33,000 mph, with a force equal to half a million tons of TNT, devastating the country over a range of at least a hundred miles. On impact, 90 per cent of the missile was vapourized; it is the remaining 10 per cent which remains buried under the south-eastern rim.

All life in the danger-zone must have been wiped out, but at that epoch there were apparently no human inhabitants. The age of the crater has been given, probably reliably, at 22,000 years. It has been suggested that some Indian legends about 'a place of fire and smoke' relate to Meteor Crater, but the time-scale is all wrong, and the suggestion is certainly out of court. The legend may well refer to Sunset Crater, which has no doubt erupted many times in the past, and will probably do so again in the future.

The remarkable preservation of the crater is, of course, due to the nature of the ground in this area. There is sandstone down to an appreciable depth, below which is a layer of what is called kaibab dolomite. Vegetation is sparse and scrubby;to be alone at the bottom of the crater — as I have been — is an awe-inspiring experience even in the daytime, and more so at night. Everything is calm, one seems to be cut off from the rest of the world.

Other impact craters are known, notably Wolf Creek in Australia, which is on the same pattern as the Arizona crater though considerably smaller. In some lists, however, we find structures which are not meteoritic — such as the Vredefort Ring in South Africa, which has been carefully studied by geologists and found to be of volcanic origin.

If a large meteorite fell upon a city or a densely-populated area, the death-toll would be colossal, and from our point of view it is fortunate that such events are so rare. During the present century there have been two major falls, both in Siberia. In 1908 a missile hit the Tunguska region, blowing pine-trees flat over a wide radius. No crater was formed; unfortunately our information is not complete, because no expedition reached the site until almost 20 years after the fall, but it seems likely that the impacting object was the icy nucleus of a small comet (suggestions that it could have been a nuclear explosion or an alien space-ship do not seem very plausible!). Then, in 1947, a giant meteorite fell in the Vladivostok area, breaking up during its descent and producing a swarm of small craters. There were no casualties on either occasion, and in fact there is no well-authenticated case of any human death due to a meteorite, though it is on record that a luckless dog in Egypt was in the wrong place at the wrong moment.

Meteorites are plentiful in the Solar System. They are quite unlike the tiny, friable objects which produce the familiar shooting-star trails, and they are more nearly related to the minor planets or asteroids. Most of the asteroids keep to a well-defined belt between the orbits of Mars and Jupiter, but there are some

The Barwell Meteorite fragment found by Patrick Moore at Barwell, Leicestershire. Weight — 1½lb Composition — stony.

which swing inward and approach the Earth; in 1937 a small asteroid, Hermes, brushed past us at less than twice the distance of the Moon. Therefore, we could unquestionably be struck, and this has certainly happened in the past, but the danger is so slight that it is no cause for alarm. You are much more likely to be run over by a bus than killed by a tumbling meteorite.

Meanwhile, Meteor Crater remains for our inspection. It is a monument of many moods; by day it is starkly beautiful, in the moonlight it seems almost gentle, but it is always peaceful, and it is hard to realize that for a few moments, more than 20,000 years ago, this was the most violent and the most dangerous place on the whole surface of the Earth.

5 SEASONS OF OTHER WORLDS

Spring, summer, autumn, winter . . . the seasons are so much part of our lives that it never occurs to us to question them. Yet how many people pause to reflect just what causes them?

The Earth's orbit round the Sun is not a perfect circle. The distance ranges between 91½ million miles in December out to 94½ million miles in June, so that considered as a whole the Earth receives rather more solar energy during our winter in England. The seasons are due not to the changing distance, but to the tilt of the Earth's axis, which is inclined to the perpendicular by $23\frac{1}{2}°$ if we reckon from the orbital plane. During northern summer, the north pole is tilted toward the Sun and the south pole away from it, while in northern winter the conditions are reversed. This effect more than compensates for the slightly lesser distance around December.

It is true, of course, that not all summers are alike; in Britain, for instance, that of 1976 was gloriously hot and sunny, while that of 1980 was drab. Unquestionably there are long-term cycles too, which affect the climate, and it is hard to doubt that the basic cause is a slight alteration in the output of solar energy, but it is impossible to be sure. Certainly we cannot put the bad summer of 1980 down to nuclear experiments or volcanic activity. Looking back at the records, it seems that in the 1680s there was a period during which the Thames froze every winter. This was in the middle of the so-called 'Maunder Minimum', when the Sun showed almost no spots (it lasted from 1645 to 1715). And there is no doubt that since really accurate records began, the very worst summer was that of 1879. However, these effects are global, not seasonal; and if the tilt of the Earth's axis were different, our seasons also would be different.

For example, consider Mars, which is further away from the Sun than we are. It has an orbital period or 'year' of 687 Earth-days, but this is equivalent to only 669 Martian days or 'sols', because Mars spins rather more slowly; its rotation period is 24 hours $37\frac{1}{2}$ minutes — slightly more than half an hour longer than our own. The axial tilt is 24°, again slightly greater than that of the Earth, but the inclinations are so nearly the same that we would expect the Martian seasons to be of the same type. This is true enough, but there are some complicating factors.

First, the Martian orbit is more eccentric than that of the Earth. The distance from the Sun ranges between $128\frac{1}{2}$ million miles at closest approach (perihelion) and $154\frac{3}{4}$ million miles at furthest recession (aphelion). All planets move at their quickest when near perihelion; and since Mars, like the Earth, has its south pole turned sunward at perihelion, the southern summers must be shorter and hotter than those of the northern hemisphere, while the southern winters are longer and colder. The distribution can be worked out easily enough:

	Length	
Martian seasons	**days**	**sols**
Southern spring (northern autumn)	146	142
Southern summer (northern winter)	160	156
Southern autumn (northern spring)	199	194
Southern winter (northern summer)	182	177
	687	669

On Earth, there is much more ocean in the southern hemisphere than in the northern, which stabilizes the temperature; this is not so for Mars, where there is no sea at all (in fact, the land surfaces of the two planets are just about equal). On Mars, the lack of water, plus the greater orbital eccentricity, means that the southern climates are considerably more extreme than those in the north. There is visual confirmation of this in the behaviour of the white polar caps, due to frozen material. The south polar cap can become considerably larger than its northern counterpart, while at minimum it is smaller.

There are clear indications of past running water on Mars; the features shown on the Mariner 9 and Viking photographs, obtained from close range, look so much like dry riverbeds that few people doubt that they *are* dry riverbeds. Yet there can be no liquid water on the Martian surface now, because the atmospheric pressure is too low: less than 10 millibars everywhere. Therefore, when running water existed, the atmosphere must have been much denser than it is today. There could be a clue here in the changing tilt of the axis, which ranges between 35° and only 13° over a period of around 100,000 years. There could be times when a polar cap disappears, releasing its frozen volatiles and thickening the atmosphere temporarily. If so, then Mars may be subject to very marked long-term climatic variations, and we may be seeing the planet at an epoch when it is at its most inhospitable. I admit that I am not too happy about this explanation, and it may be that there are periods when the Martian active volcanoes burst forth, expelling gases and vapours in sufficient quantity to thicken the atmosphere; but we do not really know, and it is true that most astronomers tend to regard Mars' huge volcanoes as extinct.

No doubt Mars will be reached by men in the foreseeable future; technically this could probably be done within 50 years, though whether it is actually achieved by then depends upon politics rather than science. Meanwhile, it is interesting to work out a possible Martian calendar. There have been plenty of suggestions. My own is to divide a Martian year into 18 months, each of 37 sols. This makes 666 sols. We need 669, so an extra sol can be tacked on to months 6, 12 and 18, giving them 38 sols each instead of 37. This should work reasonably well, though there will have to be a minor adjustment to allow for the fact that the revolution period of Mars is not exactly 687 days, but 686 days 23 hours 52 minutes 31 seconds according to Earth reckoning.

Incidentally, the north polar star of Mars is not the same as ours. We have Polaris, in Ursa Minor; Mars is better served, with the more brilliant Deneb in Cygnus, though of course Deneb will not retain this distinction indefinitely, because the Martian axis, and therefore the positions of the celestial poles, shifts. (In the distant future, Earth will have the magnificent blue Vega as its north polar star.) There is no convenient 'month' on Mars if we reckon from its two tiny satellites, Phobos and Deimos; Phobos moves round the planet in only 7½ hours, completing about three revolutions per sol, while Deimos takes 30 hours.

Venus is totally unlike Mars. Here the orbit is practically circular, with a mean distance from the Sun of 67 million miles; the revolution period is 224¾ days. But Venus is a slow spinner. The planet itself turns round only once in 243 days (though the upper clouds take a mere 4 days) and the axial inclination is 178° — so if we take the Earth as a standard, Venus spins in a wrong-way or retrograde direction. The Sun would rise toward the west and set toward the east, while the interval between successive sunrises would be 118 Earth-days. Yet from the surface of Venus, the Sun would never be seen; it would be hidden by the dense, choking atmosphere. Seasonal effects would be almost absent on Venus, and the temperature is much the same over the whole of the planet. We cannot imagine that any life of our kind could survive there.

Mercury, the innermost planet, is different again. It is much smaller than Venus or the Earth, and has virtually no atmosphere. The revolution period is 88 days, and the rotation period 58½ days, two-thirds of a Mercurian 'year'. The axial inclination is less than ours, but there are complications due to the relatively high orbital eccentricity; at aphelion Mercury is 43½ million miles from the Sun, at perihelion only 28½ million miles. There are two 'hot poles', one or the other of which will receive the full blast of solar radiation when Mercury reaches perihelion, and the apparent path of the Sun across the sky will be erratic, because of the slow rotation combined with the rapid orbital velocity. To an observer on the Mercurian equator, the Sun will remain above the horizon for 88 Earth days after having risen.

When we come to the giant outer planets, we can more or less neglect seasonal effects, because the surfaces are always so cold. Jupiter has an almost 'upright' axis, with an inclination of only 3°, the tilts of Saturn and Neptune are only slightly greater than that of the Earth (26½° and 28¾° respectively), but Uranus is a celestial oddity, with a tilt of 98° — more than a right angle, so that there are times when a pole is presented to the Sun and to the Earth. Each Uranian pole will have a 'night' lasting for 21 Earth years, and a 'day' of equal length, though for the rest of the 84-year revolution period the conditions will be less extreme. All four giants spin in less than 24 hours, so that their calendars will be complicated; but since their surfaces are gaseous, and there can be no hope of landing there, calendars are of academic interest only.

It is certainly true that the seasons on a planet must have a very marked influence upon the evolution of life there. So far as we are concerned, the Earth is eminently suitable. We do not always appreciate how lucky we are!

6 SUPERNOVÆ, PULSARS AND SCINTARS

One of the most famous of all astronomical objects is the patch of expanding gas known as the Crab Nebula, in the constellation of Taurus, the Bull. It is much too faint to be seen with the naked eye, or low-power binoculars; small telescopes show it only as a misty blur, but photographs taken with large instruments bring out an amazing complexity of detail. It was the first object in a list of star-clusters and nebulæ drawn up in 1781 by the French astronomer

Charles Messier, so that its official designation is M1. The nickname of the 'Crab' was bestowed on it by the third Earl of Rosse, when he looked at it through his giant home-made telescope more than half a century later.

The Crab is a supernova remnant — nothing more nor less than the death-throes of a formerly bright star. The actual outburst was seen in the year 1054, and for a while the supernova became brilliant enough to be visible with the naked eye in broad daylight, but then it slowly faded away. Today, we have the Crab Nebula to mark the site. Inside the expanding cloud is a tiny, super-dense object made up of neutrons (that is to say, atomic particles with no electrical change), and it is this neutron star, or 'pulsar', which is the Crab's power-house. It cannot be more than ten miles in diameter, but it is so dense that a cupful of it would weigh thousands of millions of tons; its total mass is about the same as that of our Sun.

The Crab pulsar emits radiation over a wide range of wavelengths, from the very short X-rays to the very long radio waves. The pulsar is spinning round, completing 30 rotations every second, and this is why it sends out pulses of radiation (the term 'pulsar' is a convenient contraction of the original 'pulsating star'). Apparently it sends out two beams, diametrically opposed, rather in the manner of a lighthouse, that at each revolution we receive two pulses as the Earth passes through the beams. The visible flashes are very faint, and cannot be detected by ordinary visual observation, but computer techniques have allowed them to be timed very accurately.

The second pulsar to be optically identified is in the southern constellation of Vela, the Sails; it was tracked down by astronomers at the Anglo-Australian Observatory at Siding Spring, New South Wales, which is equipped with a magnificent 153 in reflecting telescope. The Vela pulsar rotates rather more slowly than the Crab's, and is more distant (about 10,000 light-years, as against 6,000 light-years for the Crab), so that it is fainter; optically, it is one of the dimmest objects ever recorded. Here, too, we have the characteristic double flash.

Undoubtedly the Vela pulsar, is too, a supernova remnant, but other remnants do not seem to be associated with pulsars. For instance, there is the celebrated Cygnus Loop or Veil Nebula in the Swan, not far from the brilliant Deneb. Here we have disconnected arcs of nebulosity, giving the impression of an expanding bubble of gas in the interstellar medium. It must surely be the remnant of an ancient supernova, but diligent searches have failed to find any trace of a pulsar.

One of the most famous of all supernovæ blazed forth in 1572, near the W of Cassiopeia; it is always known as Tycho's Star, because it was carefully and accurately described by the great Danish astronomer Tycho Brahe. At its maximum it was brighter than Venus, but it has left no Crab-type cloud of gas; all that can be detected are a few obscure fragments of nebulosity together with a radio 'shell'. Again there is no trace of a pulsar.

Therefore, can we safely claim that all supernova outbursts are bound to produce pulsars? Previously it was thought that the answer would be 'yes', but

NGC 2403. Spiral galaxy in Camelopardalis.

today most astronomers think differently. It is, of course, possible that some pulsars are undetectable because their 'beams' miss the Earth, just as an observer will be unable to see the flashes of a distant lighthouse unless he is in the right position. On the other hand, recent work has shown that some supernovæ leave remnants of a completely different kind, and these have become known as scintars. The first to be tracked lies close to the border between Aquila and Serpens, and was originally known by the unromantic designation of W50. It is extremely faint, and the fact that it lies in a rich area of the Milky Way makes it even more troublesome to identify, but it gave every indication of being a supernova remnant, mainly because of its association with a radio source very similar to that of Tycho's Star. It is 50,000 light-years away, so that the outburst took place long before there were any astronomers on Earth to observe it.

One particularly intriguing feature was the presence of a radio-emitting star in the exact centre of the shell. This was the object which became known as a scintar, because it twinkles or scintillates at radio wavelengths. There was also an X-ray source in the area, and it occurred to Paul Murdin and David Clark, of the Royal Greenwich Observatory, that the X-ray star and the radio star might be identical. They went to Siding Spring, and used the great reflector to examine the region. They had hit upon the truth; both the radio waves and the X-rays came from an innocent-looking star which was known by its catalogue number of SS 433.

The spectrum of SS 433 turned out to be most peculiar, and also very variable. The best interpretation is that SS 433 is not a single star; it is a binary system, made up of two components moving round their common centre of gravity. One member of the pair is ejecting two streamers of gas, diametrically opposed to each other, but these jets are not fixed in position; they are behaving rather in the manner of a rotating lawn sprinkler, while the star is orbiting its companion in a period of 12 days.

The component producing the jets must be very remarkable. It is extremely massive, and there have even been the inevitable speculations that it could be a Black Hole. There is no longer any doubt about the real existence of the jets themselves, because further photographs have shown wisps of nebulosity which are caused by the jets striking the shell of the old supernova remnant.

SS 433 was the first scintar, but other have been strongly suspected. One candidate, in the far south of the sky, is Circinus X1. It also is a star which emits both X-rays and radio waves, and it is close to a supernova remnant, though no comparable jets have been identified so far. No doubt more scintars will be found in the near future.

Supernovæ are the greatest explosions known to science, and it is a pity that they are comparatively rare. In our Galaxy, the only supernovæ to have been definitely seen over the past thousand years have been those of 1006 (in Lupus), 1054 (the Crab), 1572 (Tycho's Star) and 1604 (Kepler's Star, in Ophiuchus), all of which appeared before the invention of the telescope. Fortunately, supernovæ are so luminous that they can also be detected in external galaxies, many millions of light-years away. One of them — S Andromedæ of 1885, in the famous Andromeda Spiral — even reached the fringe of naked-eye visibility. During the past few decades we have learned a great deal about the way in which they behave, but many problems remain to be solved, and it now seems that their remnants are much more diverse than used to be thought. Some produce pulsars together with clouds of expanding gas; others seem to yield gas-clouds only, while now we have the strange binary scintars to take into account. Indeed, scintars have burst upon the astronomical scene as the latest examples of objects which were not only unknown, but also unsuspected. It is fair to say that SS 433 has taken us by surprise.

7 SATURN: THEN AND NOW

Saturn, the outermost of planets known in ancient times, must have been observed at the very dawn of human history; at its best it outshines most of the stars. Almost as soon as telescopes were invented, astronomers realized that it is a most unusual object, and in the mid-17th century Christiaan Huygens correctly deduced that it is surrounded by a system of rings. That seems a long time ago; in our own time space-probes have been sent to it, and on November 12 1980 the Voyager 1 pass, at a mere 77,200 miles from the cloud-tops, marked the climax of years of planning.

Earlier in 1980 I had spent some time at the Lowell Observatory at Flagstaff in Arizona, and I was able to make use of the great 24 in refractor — which I knew well because in the pre-Apollo period I had carried out extensive Moon-mapping with it. The Lowell refractor is a famous telescope. With it, Lowell himself made all his observations of Mars. More importantly, V. M. Slipher had used it to examine the spectra of the objects we now know to be galaxies, and had found that almost all of them show Doppler red shifts, indicating that they are racing away from us. I have, I think, observed with most of the world's largest refractors, and I have yet to find one which I like better than the Lowell.

In early 1980 Saturn was well placed, and I paid considerable attention to it. This was a particularly interesting period, because the rings were edgewise on to us (as had last happened in 1966), and in smaller telescopes they vanished, because they are extremely thin. Even with the 24 in I was hard pressed to keep them in view, and they appeared only as a wafer-thin line of light. Obviously, the disk itself was excellently displayed, because no part of it was hidden by the rings, and the small inner satellites were also seen to advantage.

It is interesting to look back at the early ideas about Saturn. Galileo saw the rings with his tiny telescope in 1610, but he did not appreciate their true

Saturn, as I drew it with X400 on my 15 in reflector when the rings were edgewise-on.

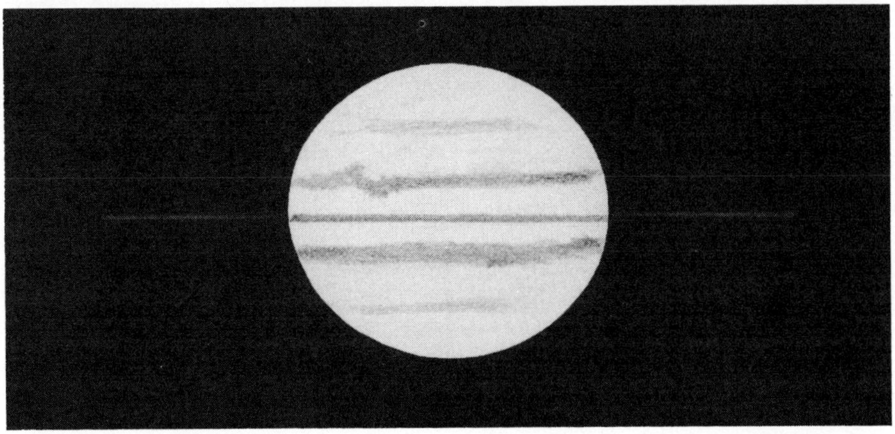

1909
ECS
1912
ECS
1911
ECS
1916
ECS
1915
ECS
1917
ECS
1920
1921
SATURN

nature, and believed that Saturn might be a triple planet. He wrote: 'Saturn is not one alone, but is composed of three, which almost touch one another and never move nor change with respect to one another. They are arranged in a line parallel to the Zodiac, and the middle one is about three times the size of the lateral ones.' Two years later he found, to his surprise, that the minor bodies had disappeared: 'Are the two lesser stars consumed after the manner of solar spots? Has Saturn, perhaps, devoured his own children? The unexpected nature of the event, the weakness of my understanding, and the fear of being mistaken, have greatly confounded me.'

What had happened, of course, was that the rings had become edge-on, so that it is no wonder that Galileo lost them. The true explanation was not forthcoming until Huygens published his book in 1659, and stated that Saturn is 'surrounded by a thin flat ring, which nowhere touches the body of the planet, and is inclined to the ecliptic'.

Oddly enough, Huygens' explanation was met with strong opposition. The French mathematician de Roberval believed that Saturn was surrounded by 'a torrid zone giving off vapours', transparent but reflective enough to produce an elongated appearance, while Honoré Fabri suggested that there were two large, dark satellites close to Saturn and two brighter ones further away.

In 1658 Sir Christopher Wren (who, remember, was a professor of astronomy long before he turned to architecture) had worked out a theory according to which Saturn had an elliptical corona, meeting the globe in two places and rotating with Saturn once in each sidereal period. However, Wren never published his theory, because before he was ready to do so he heard about Huygens' work, and accepted it immediately. Fabri was not so wise. In 1660 he published a violent attack both on Huygens' theory and upon Huygens himself. Strangely, Fabri admitted that his alleged satellite of Saturn was not really a 'moon' at all, because it did not circle round the planet! Had it done so, it would have been observed to cross in front of Saturn's disk, as well as being regularly eclipsed by the shadow of the planet. Therefore, it had to circle 'behind' Saturn all the time. (To be precise, the pamphlet was ostensibly written by one Divini, but it seems certain that Fabri was the real author.)

The pamphlet was dedicated to Prince Leopoldo, who commented that he would suspend judgement upon it until the periods of Fabri's four satellites had been found. 'That,' said Huygens dryly, 'will take a long time.'

The argument dragged on for some years, but it was settled by the improvements in telescopes, and in 1665 even Fabri had to admit that Huygens had been right. Ten years later G. D. Cassini made the first definite observation of the main gap in the rings which is still called Cassini's Division — and, until the space-probe era, was assumed to be empty, though we now know that it contains a number of very narrow ringlets.

William Herschel, perhaps the greatest of all observers, made some observa-

Opposite *Saturn 1909 to 1921. Lowell Observatory. P. Lowell and E. C. Slipher. 24 in refractor.*

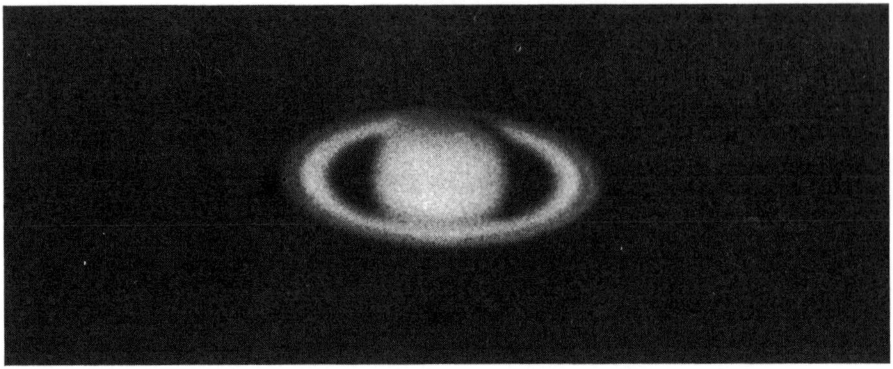

Saturn — rings open 1973 (Commander H. R. Hatfield RN).

tions of the rings between 1777 and 1781, and believed them to be as solid and substantial as the planet itself, mainly because of the shadows cast by the rings on to the globe. Later he made some further observations suggesting that as well as being flattened at the poles (which is obvious enough) Saturn was also slightly flattened at the equator, giving it a sort of square-shouldered appearance. For once Herschel was wrong, and apparently he was misled by an unusual optical arrangement involving the ring position, the duskiness of Saturn's polar regions, and the low altitude of the planet at the time. It was 1850 before Ring C, the so-called Crêpe Ring, was found, by which time Johann Encke had reported the existence of a narrow division in the outer bright ring which we know as Ring A.

What were the rings? They could not be solid or liquid, as Herschel had believed; such a ring would be torn to pieces by Saturn's strong gravitational force. As early as 1705 J. Cassini had suggested that they were made up of swarms of particles, and in 1875 James Clerk Maxwell provided a convincing mathematical proof. Twenty years later J. E. Keeler showed spectroscopically that the inner sections of the ring system move round Saturn faster than the outer parts — as was only to be expected.

So far as the planet itself was concerned, all sorts of theories were proposed. Originally it was assumed that Saturn was a kind of miniature sun, radiating enough heat to warm its system of satellites. In 1882 R. A. Proctor wrote his famous book *Saturn and its System* containing the following remarkable description:

'Regarding the cloud phenomena of the giant planets as generated by internal forces, whose real seat lies deep below the visible surface of the cloud belts, we see that these forces must be of tremendous energy, must produce enormous changes in the cloud-laden atmosphere (with effects extending widely, both vertically and laterally), and imply enormous heat in the whole frame of each planet . . . Over a region hundreds of thousands of square miles in extent, the flowing surface of the planet must be torn by sub-planetary forces. Vast masses of intensely hot vapour must be poured forth from beneath, and, rising to enormous heights, must either sweep away the enwrapping mantle of cloud which had concealed the surface, or must itself form into a mass of cloud.'

This idea was still current until 1923, when Sir Harold Jeffreys published the first of an important series of papers showing that the dwarf-sun theory was absolutely untenable. In 1926 he suggested that a more likely model involved a solid core overlaid by a thick layer of ice, which was in turn surrounded by an atmosphere amounting to about 23 per cent of Saturn's radius. Jeffreys' ideas were elaborated in 1938 by Rupert Wildt, who again worked out a rock-ice-gas model. He was followed in 1951 by W. Ramsey, whose theory assumed that Saturn was made up chiefly of hydrogen, so compressed near its core that the hydrogen started to take on the characteristics of a metal.

Today we think differently. Saturn must indeed have a solid core, and a gaseous atmosphere, but the bulk of the globe is probably liquid, with hydrogen as the main constituent — metallic at great depths, molecular higher up. And although the surface is bitterly cold, the core temperature may be of the order of 15,000°C.

Yet while we had to study Saturn from afar, we could not hope to find out the real truth. That has been left to the space-craft; first Pioneer 11, and now the Voyagers. We have come a long way since Richard Proctor wrote his book a century ago.

8 THE QUADRANTIDS

In the first days of each year we see a meteor shower — the Quadrantids. They are not always spectacular, but they are extremely interesting, and sometimes they can provide rich displays.

Meteors are tiny particles, usually smaller than pins' heads, moving round the Sun. So long as they keep well clear of the Earth they are too small to be detected; but when a meteor comes below a height of around 120 miles, it has to contend with the atmosphere. Friction is set up, and the meteor perishes in the streak of luminosity which we call a shooting-star. To be more accurate, what is seen is not the particle itself, but the effects in the air which it produces. It may be really brilliant for a brief moment; its entry speed may be as high as 45 miles per second.

Many meteors travel round the Sun in shoals, and each time the Earth passes through a shoal we see a shower of shooting-stars. Each shower has its own particular radiant — that is to say, the point in the sky from which the meteors seem to come. This is purely a question of perspective. One way to show what is meant is to picture the scene from a bridge overlooking a motorway; the lanes of the motorway are parallel, but seem to converge at a point near the horizon, which may be called the radiant of the lanes. Meteors of a shoal are moving through space in parallel paths, so that they too seem to come from a definite radiant, named generally after the constellation in which it lies. Thus the

August Perseids come from the direction of Perseus; the November Leonids from Leo, the Lion; the April Lyrids from Lyra, and so on.

Not all meteors are members of shoals. There are also sporadic meteors, which may appear from any direction at any moment. But it is the showers which are the most spectacular, and now and then we have truly magnificent displays. Thus in 1799, 1833, 1866 and again in 1966 the Leonids surpassed themselves, so that for several hours meteors seemed to 'rain down like snowflakes', though in most years the Leonids are sparse; everything depends upon whether or not we hit the thickest part of the swarm. On the other hand, the Perseids are reliable, and anyone looking up into a dark, clear sky at any time during the first fortnight in August will be unlucky not to see at least one meteor.

Meteors are associated with comets, and may be regarded as cometary debris. The Perseids move in the same path as a rather faint periodical comet, Swift-Tuttle, while the Orionids of October are associated with Halley's Comet.

It is important to note that meteors are not directly associated with the much larger bodies known as meteorites, which reach the ground without being destroyed, and may even produce craters. A meteorite is not a large meteor; it is different in nature, and meteorites are much more closely related to the minor planets or asteroids, most of which move round the Sun between the orbits of Mars and Jupiter.

The Quadrantid shower of early January takes its name from a constellation which has long since been dropped from the official lists. Ptolemy, the last great astronomer of classical times, named 48 constellations, but he could know nothing about the far southern stars, which never rose from the latitude of his home in Alexandria. Quite apart from tackling the southern regions, later astronomers made a habit of tacking on new groups by 'stealing' stars from constellations which were already accepted. One such astronomer was a German, Johann Elert Bode, who introduced groups with barbarous names such as Sceptrum Brandenburgicum (the Sceptre of Brandenburg), Lochium Funis (the Log Line) and Officina Typographica (the Printing Press). Two of his better efforts were Felis (the Cat) and Quadrans Muralis (the Mural Quadrant), but these too came under the official axe in 1934, when the International Astronomical Union revised all the constellation boundaries and pruned the list. Quadrans disappeared. It had never been really worthy of separate identity, since it had contained only some faint stars in the region of Ursa Major, the Great Bear, but it has given its name to the Quadrantid meteor shower, whose radiant lies in that part of the sky where Quadrans used to be. The radiant lies between the stars Theta Draconis and Beta Boötis, not very far from the tail of the Great Bear.

The main characteristic of the Quadrantid shower is its brief duration. The Perseids last for over a fortnight, the summer Capricornids persist for at least a month, but the Quandrantid maximum is short and sharp. The usual date is January 3 — and if the Moon happens to be near full, most of the meteors are drowned in the brilliant light.

The quick rise and fall in activity shows that the Quadrantids are 'bunched up' rather than spread all along the orbit of a comet, as the Perseids are. For that matter there is no known comet associated with the Quadrantids, but there may once have been. Mathematical investigations seem to show that the comet used to have a period of about 5½ years, and that it disintegrated between 1,300 and 1,700 years ago. Comets are relatively short-lived, and there are several well-proven cases of periodical comets which no longer exist. The classic example is Biela's Comet, which was seen in 1845 and observed to split in two. The twins returned on schedule in 1852, but have never been seen again, though in 1872 a meteor shower was observed in its place. The same fate presumably overtook the Quadrantid comet, though of course we cannot be sure.

Another interesting characteristic of the shower is that it will not last indefinitely, and may not have been seen at all before the early part of the 19th century. It was first reported on January 1 1835, though the radiant point was not accurately determined until 1863. Since then it has been consistent. Sometimes it is more spectacular than usual — as in 1977 when there were many bright meteors; I saw one or two which became almost as brilliant as the full moon. But the orbit of the swarm is still changing, and the main culprit is the giant planet Jupiter, which is extremely massive and has marked effects upon meteor streams as well as the flimsy comets. Every 11¾ years Jupiter passes very near the 'aphelion' of the Quadrantid stream (that is to say, the point at which the meteors are at their furthest from the Sun), and the long-term effect will probably be to alter the orbit so much that the shower will no longer meet the Earth. In about five centuries from now, there may be no more Quadrantids in our skies.

This will be a pity, since the present Quadrantids can be so rich. Meteor frequencies are measured in terms of what is know as the ZHR or Zenithal Hourly Rate — the number of meteors which would be seen by a naked-eye observer under ideal conditions, with the radiant at the zenith or overhead point. Of course these conditions are never attained, but the ZHR is a good general guide. For the Perseids, it is usually about 70; for the Leonids only ten, except on the rare occasions of a true 'storm'. The average ZHR of the Quadrantids is of the order of 100. The brighter meteors leave fine trains for a second or two, and the stream is extremely rich in fainter meteors.

So do not forget to look out for the Quadrantids at their best. One never knows quite what to expect, and there is something remarkably awe-inspiring about these displays of cosmical fireworks.

9 THE PALOMAR GIANT

Drive up the winding road to the top of Palomar Mountain, in the State of California, and you will find one of the most famous buildings in the world. In fact you come upon it almost without warning; it is a huge dome, as large as the

Pantheon in Rome, graceful and imposing. Inside it is the 200 in Hale reflecting telescope, whose giant 'eye' allowed astronomers to see further into space than ever before.

It is no accident that Palomar was chosen. The Observatory is well over 5,000 ft above sea level, so that the air is generally clear and transparent; the nearest city, Los Angeles, is many miles away, so that until recently its lights gave no trouble to astronomers who wanted the sky to be as dark as possible. The telescope was brought into use in 1948, and for over a quarter of a century remained not only the largest in the world, but imcomparably the largest.

It was the brainchild of a remarkable American, George Ellery Hale, who began his career as an observer of the Sun, but who knew that to study the remoter parts of the universe it is necessary to use telescopes of tremendous light-gathering power. This is not so important with bodies of our own Solar System, whose distances are measured in millions of miles; even a small telescope will show the craters of the Moon, the polar caps of Mars and the rings of Saturn. But the stars are so remote that their light, travelling at 186,000 miles per second, takes years to reach us, and there is never enough light available to satisfy the astronomer. So Hale planned reflecting telescopes with very large mirrors, and persuaded friendly millionaires to finance them — something which was much easier half a century ago than it is today!

The first of his great reflectors had a 60 in mirror; it was completed in 1908 and set up on Mount Wilson, within sight of Los Angeles. It was followed nine years later by an even larger telescope, also on Mount Wilson; this had a 100 in mirror and quickly proved its worth. But Hale was not satisfied. His call was

Left *Dome of the Palomar 200 in reflector as I photographed it in 1981.*

Right *Palomar 200 in reflector.*

always for 'More light!' and he saw no reason why there should not be a reflector with a mirror as large as 200 in in diameter. Financial backing was obtained, admittedly after seemingly endless negotiations, and construction began. Sadly, Hale himself did not live to see the telescope completed. He died in 1938, and it was not until a decade later that the giant 'eye' was first turned skyward.

It had become painfully clear that Mount Wilson was not an ideal site. The lights of Los Angeles were becoming ominously brilliant, and so the new telescope was destined for Palomar, which is much further out. All sorts of problems had to be overcome, including the replacement of the old mountain track by a proper road, and the mirror itself proved to be difficult to cast; quartz was tried, with no success, and pyrex was eventually substituted. But in the end all was well, and the story of the Hale telescope has been one of unbroken triumph.

Go into the dome, and the first thing you will see is what looks like a giant horseshoe. This is the mounting of the telescope and even though it weighs 500 tons it can be moved smoothly and with absolute precision. Inside it is mounted the telescope itself, which has no solid tube; it is of skeleton construction, with the main mirror at the bottom. The light from the object to be studied comes through the opened slit in the dome, strikes the mirror, and is reflected upward, so that it is brought to focus at the observer's 'cage' slung inside the tube itself. There is no need for a secondary mirror, as with smaller telescopes, and therefore there is no extra loss of light.

The mirror is accurately curved to within a tiny fraction of a millimetre and is coated with a very thin layer of aluminium, so as to make it as reflective as possible; every scrap of light has to be collected. Originally, the observer spent his working time inside the cage, photographing objects so far away that their light started on its journey hundreds, thousands or even millions of years ago. Things are different today, but everything still depends upon the light-grasp of the 200 in mirror.

A telescope of this size is not used to examine relatively nearby objects, such as the Moon and planets. (A few spectacular pictures of this sort were taken when the telescope was first completed, but, to be honest, they were 'for popular consumption' only.) Magnificent views have been obtained of remote star-systems, some of which are spiral in form, like Catherine-wheels, but these too form only a small part of the work carried out with the 200 in.

A telescope collects light; a spectroscope splits this light up, giving information about the nature of the light-source. Since the stars and star-systems look relatively faint, the aperture of the main telescope is all-important, and it is true to say that the 200 in was responsible for a complete revolution in outlook. Take, for example, the case of the short-period variable stars known as Cepheids. These stars do not shine steadily; they brighten and fade regularly, and the way in which they behave gives a key to their real luminosity, from which their distances can be calculated. Using the Hale reflector in the early 1950s, Walter Baade found that the Cepheids in external galaxies were much more powerful, and hence much more remote, than had been believed. It followed that the galaxies in which they lay were also more remote — and Baade showed that the universe was twice as large as expected. (I well remember the somewhat stunned silence at the meeting of the Royal Astronomical Society, in 1952, in which Baade announced his result.) Without the power of the 200 in, this fundamental discovery could not have been made.

In Hale's day, almost all astronomical work was carried out by means of photography, but nowadays electronic devices have taken over, and the 200 in has been well able to adapt to the new techniques. Moreover, the observer can carry out his work in relative comfort. It is no longer necessary to spend long, cold nights in the observer's cage. The setting and movement of the telescope is controlled from a room leading off the main dome and the results appear on a television screen. Indeed, during a typical night's observing there is nobody inside the main dome itself.

Objects in the sky send out radiations at all wavelengths; visible light makes up only a very small part of the total range. Radio telescopes, for instance, collect and analyze the very long-wavelength radiations which do not affect our eyes. In the early 1960s it was found that strong radio emissions were coming from what looked like a faint bluish star. With the 200 in, Maarten Schmidt studied the optical spectrum of the object and found that it could not be a star or a normal galaxy; it was extremely remote and almost incredibly luminous. This was the first quasar. Many others have since been found, and here again the light-grasp of the Hale reflector has been decisive.

And yet there are hints of trouble ahead. Hale chose Palomar as a site because of its remoteness, but by now Los Angeles is spreading, and so is the city of San Diego, so that the skies are no longer as dark as they used to be. Efforts are being made to minimize the trouble, both by improving astronomical techniques and by persuading the city authorities to reduce light pollution to an acceptable level. Whether these efforts will be successful remains to be seen; we can only hope so.

It is true that some smaller reflectors, such as those at Siding Spring in Australia and Kitt Peak in Arizona, are more 'modern' and more sophisticated than the 200 in, but their planners have been able to draw upon the experience gained at Palomar. In any case, there can be no doubt that the Hale reflector will remain in the forefront of research for many years to come. Never has one telescope told mankind so much in so short a time.

10 INSIDE THE HUNTER'S SWORD

There can be little doubt that Orion, the Hunter, is one of the most glorious of all the constellations. It dominates the evening sky throughout winter and early spring, and it cannot be mistaken; it has a characteristic shape, with three bright stars making up the belt, and two exceptionally brilliant stars, Betelgeux and Rigel. Rigel is a true celestial search-light, about 60,000 times as luminous as the Sun, and so remote that its light takes some 900 years to reach us; we see it today not as it is now, but as it used to be in the time of William the Conqueror. Betelgeux is less powerful but it is a huge Red Supergiant, large enough to swallow up the whole path of the Earth round the Sun.

Below the belt lies the Hunter's Sword. Here even the naked-eye observer can see a misty patch of light; we know it as the Great Nebula in Orion, or Messier 42 (because it was the 42nd object in a famous catalogue of clusters and nebulæ drawn up by the French astronomer Charles Messier two centuries ago). Apparently it was first described in the year 1610 by an otherwise undistinguished astronomer named Nicholas Peiresc. Telescopically it appears as a mass of gas, showing considerable detail, and containing the multiple star Theta Orionis, nicknamed 'the Trapezium' because of the arrangement of its four main components.

Gaseous nebulæ are contained in our own star-system or Galaxy. Some of them shine merely by reflecting the light of stars in or near them; others (such as M42) are excited to a certain amount of self-luminosity by the presence of extremely hot stars. But suppose there are no suitable stars available? In this case the nebulosity will be dark and will be detectable only because it blots out the light of stars beyond. There is no essential difference between a bright nebula and a dark one; everything depends upon the angle from which we see it.

Nebulæ are stellar birthplaces. It is here that fresh stars are being created out of the nebular material, and there is no doubt that the process is now going on inside M42, more than 1,400 light-years away from us. There are also many youthful stars which have not yet settled down to a stable state, and are still varying irregularly in brightness.

Photographs taken with large telescopes show vivid colours in the Nebula; there are for instance red areas, due to hydrogen, and green areas, due to oxygen. Unfortunately these colours are too fugitive to be seen directly by an observer at the eye-end of a telescope, because the intensity of the illumination is too low. Amateur astronomers never tire of looking at the Orion Nebula, with its filmy gas and the four stars of the Trapezium. But recently there have been some fascinating developments which involve not visual observations, but studies in infra-red.

Visible light makes up only a small part of the total range of wavelengths, or 'electromagnetic spectrum'. To the short-wave end of violet light we have ultra-violet, X-rays, and the very penetrating gamma-rays. To the long-wave end, beyond red light, we have infra-red, microwaves and then radio waves. It is these long wavelengths which are of special interest with the Orion Nebula, because they can slice through the gas and dust, whereas visible light cannot.

In the 1960s two astronomers of the California Institute of Technology, Becklin and Neugebauer, were carrying out an infra-red survey from the Mount Wilson Observatory when they discovered a very strong source deep inside the Orion Nebula. Nothing could be seen visually, but in infra-red this Becklin-Neugebauer or BN object is one of the brightest sources in the entire sky — much brighter than any of the stars in the Trapezium. This particular survey was carried out at a wavelength of 1.5 micrometres, which is three times longer than the wavelength of visible light. A few years later techniques had advanced to the point where investigations could be carried out further in the infra-red — at wavelengths over 200 times that of visible light — and another remarkable source was found; it is known as the Kleinmann-Low or KL object, after its two discoverers. It is close to BN, but is not identical with it.

Moreover, it has been found that the luminous Orion Nebula is only part of a much more extensive mass of nebulosity — it has been described as 'the tip of an interstellar iceberg' because it is very cold (around -260°C). Hydrogen is the most abundant constituent, but in this state hydrogen is shy about showing itself and has to be detected by less straightforward methods. The hydrogen acts upon molecules of carbon monoxide, which are much easier to locate; therefore, where we find carbon monoxide, we may be sure that there is hydrogen too.

What are the natures of BN and KL? We must admit that we cannot be sure, but most astronomers now think that BN is simply a very large and powerful star which is permanently hidden from us by the intervening nebulosity. KL is different. Its temperature is of the order of -170°C, and there seems to be material near it moving at very high velocities of up to 17,000 miles per hour. This does not seem to fit in with the idea of a very large, massive star. It could well be a 'proto-star'; that is to, say a star which is still in the process of formation. We may even be witnessing a stellar birth.

According to modern theory, a star begins its career by condensing gravitationally out of nebular material. As it shrinks, its centre heats up and eventually

Orion and his retinue. The Great Nebula is clearly shown.

the temperature becomes so high that nuclear reactions are triggered off. At this stage the star blows out material in all directions in a kind of stellar wind — or, more accurately, stellar hurricane. Where the flow speed reaches the speed of sound, a shock front is formed and this could be so in the case of KL.

This is only one possible explanation. A very quickly-spinning star could produce similar effects, but some very recent investigations do seem to favour the idea of a stellar infant. Inside the KL source there are several distinct centres of high-velocity expansion, in which case we are dealing not with a single proto-star, but with a whole cluster of proto-stars.

Will we ever be able to have direct views of BN and KL? The answer seems to

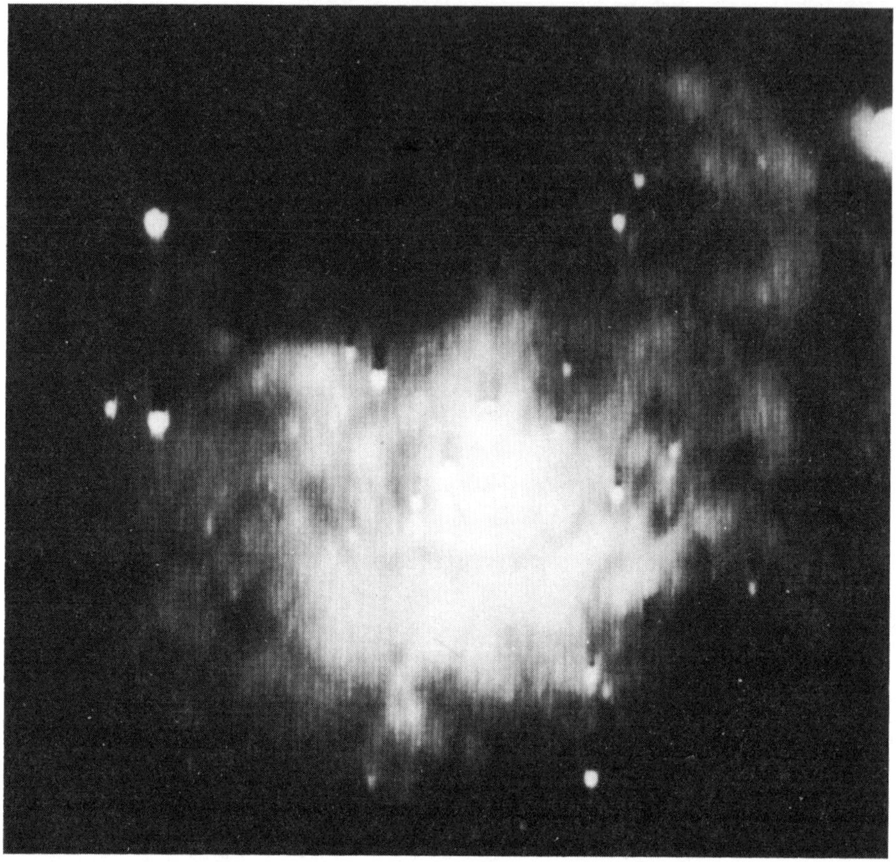

The Trapezium (Theta Orionis). One of the early pictures taken with the INT from La Palma, in February 1984.

be 'No'. We have to depend upon what is now called invisible astronomy, and it is worth noting that radio astronomy techniques and infra-red techniques are starting to overlap. Sheer visual observation is limited; it is rather like trying to see the dial of Big Ben from Parliament Square on a very foggy night.

We have come a long way since Nicholas Peiresc first drew attention to the misty patch below the Hunter's Belt. We are at least sure that we are seeing a place where stars are being born, and KL may be of special significance. Protostars develop very quickly on the cosmical time-scale; they change dramatically in a few hundreds or thousands of years and it may well be that we will eventually be able to follow the evolution of KL, assuming that it really is a star (or stars) in extreme youth. One day, perhaps, we may find out what is really going on deep inside the Sword of Orion.

11 MR HERSCHEL'S PLANET

Go to Bath Spa, and you will have no difficulty in finding New King Street. Admittedly it lies in one of the less fashionable parts of the city and does not seem to be impressive; but one of the houses in its terrace, No 19, has an honoured place in the history of science. It was from here, on March 13 1781, that a then unknown amateur astronomer named William Herschel discovered a new planet, thereby doubling the size of the known Solar System.

Herschel was not looking for a new member of the Sun's family, and did not immediately recognize its nature even when he had found it. He mistook it for a comet. This would have been interesting, but no more; faint comets are common enough. It was only when the orbit was worked out that the immense significance of the new object was appreciated. Today we call it Uranus; it is a giant planet over 30,000 miles in diameter, moving far beyond the path of Saturn, the outermost of the planets known in ancient times. It takes 84 years to complete one journey round the Sun.

Friedrich Wilhelm Herschel was Hanoverian by birth. He came of a musical family and at the age of fourteen joined the Hanoverian Foot Guards as a bandsman. The Seven Years' War was raging; young Herschel spent some months in England, then united with Hanover under King George III, after which he returned to the Continent to take part in active service under the Duke of Cumberland. In July 1757 the Duke's army was utterly routed at the Battle of Hastenbeck, and Herschel related how he had to spend a night in a water-filled ditch! Military service did not appeal to him; he had been under age when he had joined the Army, and had never been formally enlisted, so that he was able to obtain his release. A later story that he deserted is quite unfounded. His certificate of honourable discharge, signed by General A. F. von Spörken, is on show at the Herschel Museum at No. 19 New King Street.

Herschel came back to England to try his luck at a musical career. He never returned to live in Hanover, and henceforth became 'William' instead of Wilhelm. After several posts in North England he came to Bath and was appointed organist at the new Octagon Chapel as well as playing in the famous Linley orchestra which performed daily in the Pump Room.

The Chapel was opened in 1767, and Herschel was an undoubted success; his musical skill, together with his pleasant and genial personality, made him popular everywhere and he was also in great demand as a music teacher. But already a new interest was stirring; Herschel was fascinated by astronomy. He read books and borrowed a small reflecting telescope; when this was not satisfactory he decided to make telescopes for himself.

He had had no scientific training — he was completely self taught — and mirror-making is no easy task. Modern mirrors are made of glass, coated with a reflecting layer of aluminium, but in those days it was impossible to cast glass mirrors of any size and the solution was to use an alloy of copper and tin, known as speculum metal. By now he was installed at No 7 New King Street, and it was here that his first mirrors were cast. His sister Caroline joined him from

Above left *19 New King Street.* **Above right** *Wedgwood representation of William Herschel* (Josiah Wedgwood & Sons Ltd).

Hanover, both to act as his housekeeper and to follow up her own career as a singer, but before long she found herself pressed into the rôle of assistant astronomer. Without her, William's life would have been much more difficult. Caroline was devoted to him; she acted as his 'recorder', noting down all his observations as he made them, and in her own right she discovered no less than eight comets. Even William's later marriage caused no more than a temporary coolness.

Herschel did not have instant success. Failure followed failure, but at last he produced a tolerably good mirror 5 in in diameter, with a focal length of $5\frac{1}{2}$ ft. On March 4 1774 this new instrument was turned toward the Great Nebula in Orion, and Herschel's real career was well and truly launched.

In the summer of 1774 William and Caroline moved away from No 7 but in September 1777 they came back to New King Street, this time to No 19. Then, for some reason or other, they went away once more but returned to No 19 in early March 1781. William set up his new telescope, which had a mirror of 6.2 in aperture and a focal length of 7 ft, and continued with what he called his 'review of the heavens'. On March 13 he came across something which he knew at once could not be a star. It showed a disk, which no star can do, and it moved slowly from one night to another, which again no star can do. Herschel notified the Astronomer Royal, Nevil Maskelyne, and also the Rev T. Hornsby, Radcliffe Observer at Oxford. Confirmation was quick; Maskelyne at least seems to have been convinced almost at once that the new object was a planet. An orbit was worked out by the Finnish mathematician Anders Lexell, putting the

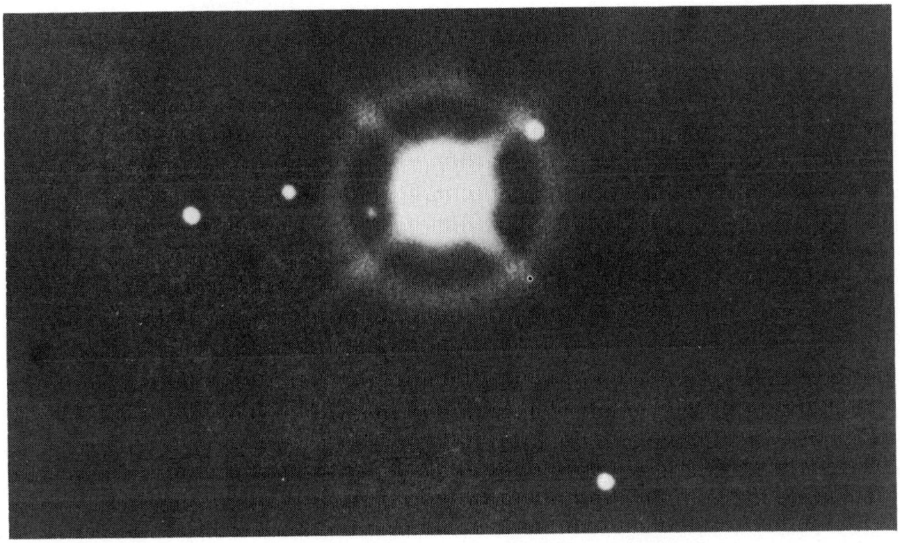

Uranus and its five satellites, photographed in 1948 by G. P. Kuiper, McDonald Observatory, Texas.

matter beyond all doubt, and Herschel's fame spread throughout the world.

Ironically, Caroline was not with him on that fateful night. She had stayed behind in the former house (in Rivers Street) to finish clearing-up. But at least we have William's own account of the discovery.

King George III, himself keenly interested in astronomy, was anxious to see this fellow-Hanoverian who had added a new member to the Sun's family, and it was this which led to Herschel's eventual departure from Bath to become the King's private astronomer at a salary of £200 per year — modest by today's standards, but enough to allow Herschel to abandon music as a career and devote his whole life to the skies. He set up a workshop in the cellar of No 19 and set to work, helped by his brother Alexander as well as by Caroline. A small furnace was built, and the materials for a proposed 36 in mirror were carefully selected. The mirror was to be cast in a mould made of horse dung after which it could be figured to the correct shape. Unfortunately there were several major hitches. Once there was a leak in the mould; molten metal ran out on to the flagstones which promptly exploded. Caroline related that 'both my brothers and the caster and his men were obliged to run out at opposite doors, for the stone flooring (which ought to have been taken up) flew about in all directions as high as the ceiling. My poor brother fell exhausted by heat and exertion on to a heap of brickbats'.

That, for the moment, put an end to mirror-making, but already Herschel had become the best telescope-maker of his time and when he took some of his instruments to the Royal Observatory at Greenwich it was clear that they were

Above *A telescope made by William Herschel — identical with that used for the discovery of Uranus.*

Left *Garden of 19 New King Street, with Patrick Moore in 1979, before restoration started.*

better than any previously built. His appointment to the King meant that he had to move nearer to Windsor; he and Caroline went first to Datchet and then to Slough, where they installed themselves in their new home, Observatory House. It was from here that Herschel's finest work was done. He made an excellent 18 in reflector with a focal length of 20 ft and went on to construct a real giant, with a 48 in mirror and a focal length of 40 ft. This telescope remained the world's largest until the Earl of Rosse built his 72 in in 1845. Frankly the 48 in telescope never really came up to expectations, mainly because it was so awkward to handle, but Herschel did make one discovery with it; that of the two inner satellites of Saturn, Mimas and Enceladus, soon after the telescope was completed in 1789.

His enthusiasm was unquenchable. He discovered thousands of double stars, clusters and nebulæ; he proved that most double stars are physically-associated or binary systems, and he was the first to draw up a reasonably accurate picture of the shape of our star-system or Galaxy. Without doubt he was the greatest observer who has ever lived, and before his death in 1822 he had received every honour that the scientific world could bestow. His son, John, followed up his father's work by extending the 'review of the heavens' to the southern skies; during the 1830s he took the 20 ft reflector to the Cape of Good Hope, and stayed there for some years, so laying the real foundations of southern-hemisphere astronomy.

Caroline returned to Hanover after William's death. The great 40 ft reflector was used for the last time in 1815; later it was dismounted and the giant tube lay in the garden of Observatory House until it was shattered by a falling tree. The surviving section of it is now to be seen in the grounds of the Old Royal Observatory in Greenwich Park. Observatory House itself was pulled down in 1960, and the site is now occupied by the Rank Xerox offices, though a monument has been erected on the site of the 40 ft telescope.

This leaves No 19 New King Street as the only 'Herschel house' still standing. For years it was neglected, but in 1976 the first steps were taken to preserve it. The William Herschel Society was founded, the house was acquired through the generosity of Dr and Mrs Leslie Hilliard and turned into a proper museum open to the public at selected times. It was officially opened on March 13 1981, exactly 200 years after Herschel made his great discovery, and I was indeed honoured at being asked to perform the opening ceremony.

So long as humanity lasts. William Herschel will never be forgotten.

12 NEPTUNE: THE OUTERMOST GIANT

The discovery of Uranus, by Herschel in 1781, took the astronomical world by surprise; but this was only the beginning of the story. It soon became clear that Uranus was not moving as it might have been expected to do. Over the years it wandered away from its predicted position; something was pulling on it, and there seemed every reason to suppose that this something was another planet, moving in an orbit still further away from the Sun. Astronomers were presented with what might have been termed a cosmical detective problem. They could see the victim, Uranus, and they had to find out just where the culprit might be.

Various suggestions were made, but the first mathematicians to tackle the problem in detail were John Couch Adams in England and Urbain Le Verrier in France. They worked independently, and knew nothing about each other. Adams, a young Cambridge graduate, finished his calculations in 1845; he was confident that he knew the position of the planet fairly accurately and he contacted the then Astronomer Royal, George Airy, in the hope that a search would be put in hand.

A chapter of accidents followed. Airy was not initially enthusiastic; Adams failed to reply to a question which the Astronomer Royal put to him, and nothing was done. Adams also contacted a well-known amateur, William Lassell, but when the letter arrived Lassell was hors de combat with a sprained ankle and the letter was mislaid. Meanwhile Le Verrier had finished his work, and after failing to interest his French colleagues he sent his results to the Berlin Observatory. Johann Galle and his assistant Heinrich D'Arrest lost no time. Using the fine Berlin refractor (now, incidentally, on display in the Munich

Museum) they set to work and found the planet almost immediately, very close to the position which Le Verrier had given. After some discussion, the new world was named Neptune.

By now Airy had heard of Le Verrier's work, and had asked James Challis, professor of astronomy at Cambridge, to begin a search. Challis had no good sky-maps of the area; he adopted a slow and cumbersome method of procedure, and he failed to compare his observations promptly. Only when Le Verrier's triumph had been announced did Challis realize that he had seen Neptune twice without recognizing it as the expected planet!

A somewhat undignified dispute about priority followed. Today Le Verrier and Adams are recognized as co-discoverers of Neptune (except by the French!), though in fact the first to identify it were Galle and D'Arrest. Since Adams was confident of the accuracy of his prediction, it is surprising that he did not undertake a search on his own account. Neptune is easily visible with binoculars (its magnitude is 7.7) and although it is slow-moving, its gradual shift against the starry background soon betrays its true nature.

Neptune; Triton and Nereid are arrowed. The image of Neptune itself is necessarily overexposed—hence the spikes!

Lassell, one imagines, was distressed by the turn of events; had he read Adams' letter he would certainly have located the planet. However, he did discover Triton, Neptune's main satellite, a few weeks after Neptune itself had been identified in 1846.

Though both Le Verrier and Adams had given remarkably correct positions for Neptune, the orbits they had predicted were very wide of the mark. In particular, Neptune was much closer to the Sun than had been expected. There is a famous mathematical relationship, known as Bode's Law, linking the distances of the various planets from the Sun; it works quite well out to Uranus, but breaks down completely for Neptune, and it is probably nothing more than coincidence. Neptune's mean distance from the Sun is 2,783 million miles; it takes 164¾ years to complete one revolution. If it were possible to drive straight from the Earth to Neptune, taking the shortest route and keeping up a steady 60 mph, the journey would take nearly 5,200 years.

Neptune and Uranus are often regarded as twins. Certainly they are similar in size and mass; Uranus is slightly the larger of the two, with a diameter of around 32,000 miles as against 30,000 for Neptune, but it is not so massive or so dense. It would take 14 Earths to equal the mass of Uranus, but 17 Earths to match Neptune.

There is another interesting difference between the two. Uranus has an extraordinary axial tilt. Instead of being inclined to the perpendicular to the orbit by about 23½°, as with the Earth, the angle is 98°, so that at times Uranus has one pole facing the Sun. With Neptune, the tilt is 29°, which is much more normal. Both planets have quick rotations. That of Neptune is now believed to be about 17 hours 40 minutes, so that there are over 80,000 Neptunian 'days' in every Neptunian 'year'.

No telescopes will show any definite detail upon Neptune's tiny, bluish disk, but spectroscopic observations have confirmed that the main constituent of the visible gas-layer is hydrogen, together with methane (which is, of course, a hydrogen compound) and helium. We are still not sure about the make-up of the globe, but probably there is a relatively small silicate core surrounded by liquid in which water and ammonia are important constituents. It is quite definite that, as a pair, Uranus and Neptune are markedly different from Jupiter and Saturn.

Perhaps the most notable difference between the two outer giants is that according to all the evidence, Neptune has an internal heat-source while Uranus has not. The reason for this is unknown. Whether Uranus' peculiar axial tilt has anything to do with it remains to be seen. Also, the atmosphere of Uranus is clear and transparent to great depths, while that of Neptune contains layers of aerosols which are decidedly variable.

Uranus has five known satellites, all smaller than our Moon. Neptune has two, possibly three, but only one of these, Triton, is large; its diameter may be of the order of 3,000 miles, so that it is comparable with Mercury. There is strong evidence that, like Titan in Saturn's system, it has an atmosphere. It is the only large planetary satellite to move round its primary in a retrograde or

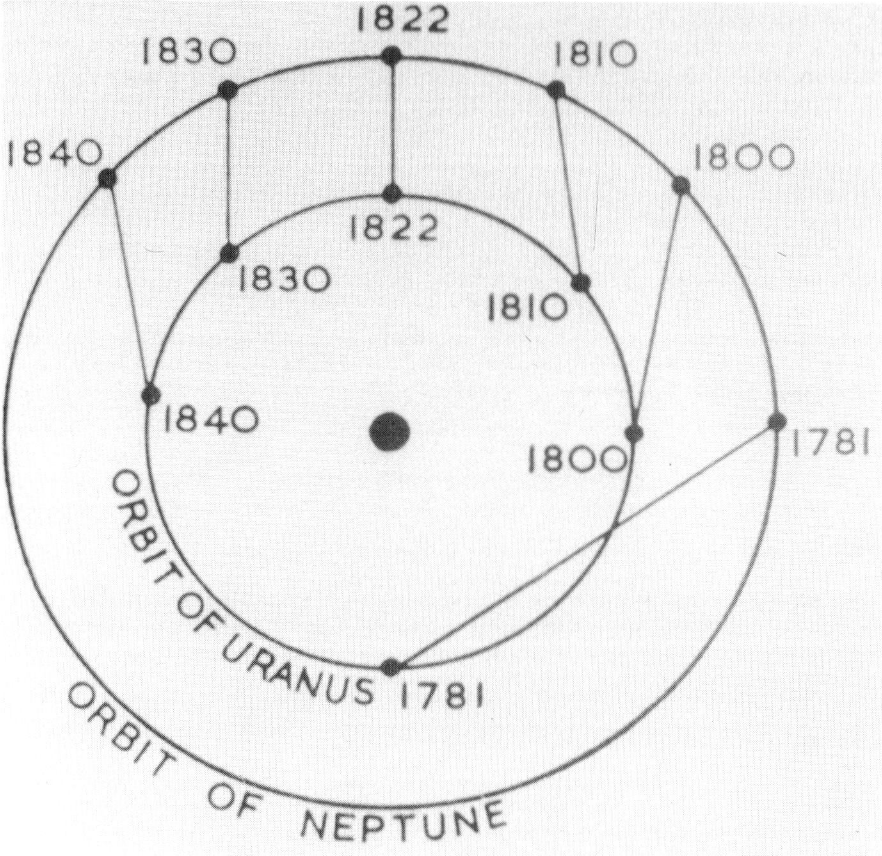

Diagram showing how Neptune pulled upon Uranus between 1780 and 1840.

wrong-way direction — that is to say, opposite to the sense in which Neptune is rotating. The other satellite whose existence has been definitely established, Nereid, is no more than 300 miles in diameter, and has a very eccentric orbit, more like that of a comet than a satellite but unlike Triton, it has direct motion.

Does Neptune have a ring? On the face of it the answer might well be 'probably'. Saturn's glorious ring-system is in a class of its own, but both Jupiter and Uranus have obscure rings and Neptune might be expected to follow the general pattern. On the other hand, the retrograde movement of Triton may make conditions above the Neptunian cloud-tops distinctly unstable and I am prepared to forecast that Neptune, alone of the giants, will prove to be ringless (though suspicions of a ring were, admittedly, announced very recently). The question may be cleared up in 1989. In that year the Voyager 2 space-craft is scheduled to make a fly-by of the planet, and send back information from close range. If it does so, it will be the first probe to rendezvous with all four giant planets in turn.

The ninth planet, Pluto, was discovered in 1930. It is now known to be smaller than our Moon, and much smaller than Triton, so that its planetary status is in doubt. Speculations that it used to be a satellite of Neptune, and broke free, rest on very uncertain evidence, and the whole idea has been weakened by the revelation that Pluto is accompanied by another small world, Charon; but certainly the retrograde orbit of Triton may indicate that some exceptional event took place in the remote past. For the moment we have to admit that there is much to be learned. The Hubble Space Telescope, due for launching before long now, may tell us more but our main hopes rest with Voyager 2. If it is still transmitting in 1989, it should give us our first really reliable news about Neptune and its system.

13 ORDEAL BY ASTEROID

'A huge cosmic boulder, nearly a mile wide and weighing over four million tons, is hurtling toward us. It is on a collision course with the Earth, and will strike us within a matter of weeks. The damage will be colossal; our only course is to attack the boulder, either by destroying it with nuclear bombs or else by diverting it from its path and making it miss the Earth. Humanity is in mortal danger!'

Science fiction? Fortunately, yes; but such a disaster is not impossible. Large boulders do wander about the Solar System, and we are not immune from impacts by them. As we have found, Meteor Crater in Arizona was formed when a missile hit the desert 22,000 years ago. Yet the damage then was comparatively local and this was also true of the Siberian impact of 1908, though it was lucky that the missile landed in uninhabited territory (actually, it may have been part of a comet, and mainly icy). A direct hit from a minor planet or asteroid would be much more devastating.

Icarus, which is about a mile in diameter, by-passed us in 1968 at a mere four million miles, and there have been closer approaches than this. The present record-holder is Hermes, which made its pass in October 1937; at its minimum distance it was only 485,000 miles away, less than twice the distance of the Moon. (Sensational newspaper headlines came out when the story was officially released, some months later; one in the London *Daily Mirror* read: 'WORLD DISASTER MISSED BY FOUR HOURS AS TINY PLANET HURTLES PAST'.) Since then many other 'Earth-grazers' have been observed, and it seems that they are more common than used to be thought.

All of them are small. The larger asteroids keep safely out of our way; they are confined to a zone in between the orbits of Mars and Jupiter, well over 150 million miles from the Sun. The senior member of the swarm, Ceres, is only 650 miles across, but most are far smaller, and only one asteroid, Vesta, is ever visible with the naked eye. Their orbits are perfectly stable, and we are in no danger from them. It was once believed that they represented the débris from an old planet or planets which met with disaster in the remote past, but it is now believed that they never made up a larger body. All the asteroids combined would not add up to one object the mass of our Moon.

The asteroids with eccentric orbits, swinging them well inside the main zone and into the inner reaches of the Solar System, are cosmic midgets. They may possibly be ex-comets which have lost all their volatiles (though personally I have reservations about this intriguing idea). They are of three types: The Amor asteroids have orbits which bring them inside the path of Mars, but outside that of the Earth; the Apollo asteroids cross the Earth's orbit; and those of the Aten group mainly keep closer to the Sun than we are. Icarus has a perihelion which carries it within the orbit of Mercury; it is then only 16,000,000 miles from the Sun, and must be red-hot, though at aphelion it moves out beyond Mars, and is bitterly cold. Certainly its climate is extremely uncomfortable.*

In the foreseeable future, Icarus cannot collide with us; but its orbit is subject to marked long-term changes, and of course it is only one of many 'Earth-grazers'. So what could we do if it threatened us? Students of the Massachusetts Institute of Technology have carried out a detailed investigation, and have come to the conclusion that a nuclear attack would be feasible. One problem would be the lack of available time. An Icarus-sized asteroid might not be detected until a few weeks or even days before collision.

Let us suppose that the impact did occur; what would the results be? If Icarus landed on dry land, it would produce a crater from 10 to 15 miles in diameter, but as the energy involved would be of the order of half a million megatons of TNT the devastated area would be much more extensive than this. An impact in, say, Birmingham would leave little of Newcastle to the north or Plymouth to the south. The outburst would be at least four times more violent than that of Krakatoa in 1883. And if the impact were in the ocean, waves in the region of 100 ft would cause world-wide damage.

Nothing like this has occurred in historical times. There are various impact craters, quite apart from that in Arizona; Wolf Creek in Australia is a good example, though one must be wary of jumping to conclusions; note, for instance, that the Vredefort Ring near Pretoria is often listed as an impact crater, though geologists who have carried out long-term on-the-spot investigations are unanimous in saying that it is volcanic. Moreover, missiles of the Arizona or Wolf Creek type are hardly to be compared with Icarus or Hermes, and are better regarded as large meteorites rather than small asteroids. (This may, of course, be a meaningless distinction, because there could well be a complete range of size between Ceres at the top of the scale down to pocket-sized meteorites at the other.)

* Until 1983 Icarus was regarded as unique, but in that year another midget asteroid, now known as Phæthon, was discovered by the equipment on board IRAS, the Infra-Red Astronomical Satellite. Phæthon approaches the Sun to within 13 million miles, much closer even than Icarus.

We must also bear in mind the surface of the Moon, which is pitted with craters of various sizes. Some are well over 100 miles across, while others are tiny. Undoubtedly some of them are due to meteoritic impact, and many astronomers believe that this was the main process in the moulding of the lunar surface features, though others (such as myself) maintain that internal forces were the more important. But whichever theory is correct, the existence of many impact craters on the Moon is not in doubt. There are also craters on Mercury, Mars and the two tiny Martian satellites Phobos and Deimos, which may possibly be ex-asteroids which wandered close to Mars long ago and were captured.

A fascinating theory has been put forward by Sir Fred Hoyle, who has been looking into the causes of the Earth's Ice Ages (the last of which ended a mere 10,000 years ago) and has come to the conclusion that they have been produced by major impacts. According to this theory, a giant meteorite will throw material into the upper atmosphere and this material will reflect the Sun's rays away into space, so that the surface temperature of the Earth will drop with alarming rapidity. Hoyle points out that there was an abrupt change in surface conditions about 65,000 years ago, when many species of living things died out — including the dinosaurs, which had been lords of the Earth for so long. The impact of a vast iron missile, Hoyle says, could have been responsible.

If cosmical collisions are indeed responsible for Ice Ages, we must agree that a new crisis might be upon us at any moment. Hoyle's remedy is to build apparatus to 'stir' the oceans, preventing the accumulation of very cold water near the sea-beds and keeping the oceans warm enough to 'store' heat and prevent the ice from spreading. Whether such a project is possible must be left to the engineers to decide. But Hoyle is certainly right when he claims that major impacts must occur in the future, just as they must have done in the past.

It is only too easy to be alarmist. The chances of an asteroidal impact are not high; they are far less than the possible outbreak of a nuclear war which would destroy civilization for ever. But the chances are not zero, and it is interesting to speculate as to what would happen if an Icarus-type missile were seen to be heading straight for Earth. One would like to think that international disputes would be shelved, and that the scientists of all countries would band together in an attempt to avoid catastrophe. This would probably not happen at the present time; but with luck the next great impact will not come until the people of Earth have learned how to live together in peace.

14 THE SUMMER TRIANGLE

From late spring until well into autumn (in the Earth's northern hemisphere, that is to say) the evening night sky is dominated by three bright stars — Vega, Altair and Deneb. Years ago I referred to this trio as 'the Summer Triangle' during a *The Sky at Night* programme and the nickname now seems to have passed into general use, though it is completely unofficial and the members of the Triangle are not even in the same constellation. Vega is in Lyra, the Lyre or Harp; Deneb in Cygnus, the Swan; and Altair in Aquila, the Eagle.

All three are of the first magnitude, and Vega is actually the third brightest star ever visible from Britain; it is surpassed only by Sirius and (just) by Arcturus. As most people know, the magnitude scale is a measure of a star's apparent brilliancy and works rather in the manner of a golfer's handicap, with the brighter performers having the lower values. Thus Vega (magnitude 0.0) is brighter than Altair (0.8), while Altair is brighter than Deneb (1.3). On the same scale Venus has a magnitude of about -4.4, while the Pole Star is a modest 2.0.

It is important to note that a star's apparent magnitude is not a reliable key to its real luminosity, because the stars are at very different distances from us. This is well shown by the three members of the Summer Triangle. Deneb, which looks the faintest of the trio, is by far the most powerful. It is well over a thousand times more luminous than Vega, and about six thousand times more luminous than Altair.

Vega cannot be mistaken, partly because of its brilliance, partly because of its position and partly because of its colour. During summer evenings it is almost at the zenith or overhead point — a position occupied during winter evenings by another star of almost exactly the same magnitude, Capella in Auriga, the Charioteer. Vega is decidedly blue, and for this reason alone it is one of the most beautiful of all the stars.

A star's colour depends upon its surface temperature. Our Sun is a yellow star, with a modest surface temperature of almost 6,000°C. Vega has a surface more than twice as hot, and this explains its blueness, Moreover, it is decidedly energetic, even though it pales before Deneb. It would take 55 Suns put together to make one star as powerful as Vega. Yet Vega appears as nothing more than a dot of light, with an apparent diameter too small to be measured directly. Therefore it must clearly be a long way away, and the distance has been found to be 27 light-years. This means that we are seeing Vega not as it is today, but as it used to be 27 years ago. When we look out into space, we are also looking backwards in time.

Both Vega and Deneb are so far north in the sky that they never actually set over Britain, though at times they almost graze the horizon. Altair, the third member of the Triangle, is closer to the celestial equator, and is out of view for part of the year. It is one of the closest of the bright stars, at a distance of only 16 light-years, and its luminosity is a mere 10 times that of the Sun. It is white, with a surface nearly as hot as that of Vega, and it is flanked to either side by a fainter star. The upper or northern star is Tarazed, or Gamma Aquilæ; the lower or southern is Alshain, or Beta Aquilæ.

It is worth looking at Tarazed, preferably with binoculars or a wide-field telescope. It is not blue like Vega, or white like Altair; it is decidedly orange, and its surface is at a temperature of less than 4,000°C. But to compensate for this it is very large and its luminosity is some 800 times that of the Sun. Tarazed is 340 light-years away.

This shows yet again that what we call a 'constellation' is nothing more than a line-of-sight effect. Altair and Tarazed look close together in the sky, and are put into the same group — Aquila; but there is no connection between the two, and Tarazed is a great deal further away from Altair than we are. From our vantage point in the Galaxy Altair appears in the foreground, Tarazed in the background. The same situation holds good for all the other constellations — which again shows up the ridiculous claims of astrologers, who believe that the constellations really mean something!

Altair is not unique in having a fairly prominent star to either side of it. Antares, in the Scorpion, is another; but there is no danger of confusing the two, because Antares is much lower in the sky, and is visible for only a limited period during summer evenings. Moreover Antares is fiery red; its name really means 'the rival of Mars'.

The third member of the Summer Triangle, Deneb, is a particularly interesting star. To the naked eye there is nothing remarkable about it; it is white, and obviously fainter than either Vega or Altair. Yet it ranks as a cosmic searchlight. It is a supergiant star, 60,000 times more luminous than the Sun, and so remote that its light takes at least 1,600 years to reach us. We are seeing Deneb as it used to be in the days of the Roman Occupation.

Stars are not 'burning' in the conventional meaning of the term. They are producing their energy by nuclear reactions. The lightest of all the elements, and by far the most plentiful in the universe, is hydrogen. Deep inside a star such as the Sun, the hydrogen is being converted into helium; it takes four 'bits' of hydrogen to make one 'bit' of helium; each time this happens a little energy is set free and a little mass is lost. It is this energy which keeps the Sun shining, and the mass-loss amounts to four million tons per second.

Furious though this rate may seem, it is not much when we remember the Sun's tremendous mass. Indeed, we know the Sun to be at least 5,000 million years old, and nothing dramatic will happen to it until its supply of available hydrogen 'fuel' begins to run low, which will not be for another 5,000 million years yet.

With Deneb, things are different. It is squandering its resources much more quickly, and it will not be able to maintain its output for nearly so long as the Sun; neither is it anything like so old. The age of Vega is perhaps 230 million years, and Altair 900 million years, but Deneb is far younger than either.

If we could look at our Sun from the distance of Altair we would see it as a moderately bright star; but from Vega the Sun would be too faint to be seen at all prominently. From the immense distance of Deneb, the Sun would fade into invisibility even with extremely powerful telescopes.

There is plenty of variety among the stars, and the members of the Summer Triangle are particularly interesting because they are so unlike each other. As they shine down from the blackness of the sky, it requires an effort of the imagination to realize that they are giant suns many light-years away.

15 ENCOUNTER WITH SATURN

On November 12 1980 the unmanned American space-craft Voyager 1 made its closest approach to the planet Saturn. Moving at over 56,000 miles per hour relative to Saturn, it passed only 77,200 miles above the cloud-tops; and though the distance from Earth was almost 950 million miles, Voyager was obeying its commands as promptly as ever. During the weeks before and after that closest approach, astronomers learned more about Saturn than had been possible throughout the whole of human history. Surprise followed surprise; each day brought its quota of new, unexpected discoveries not only about the planet itself, but also about its family of satellites. In Mission Control (where I was, throughout the encounter) some of the revelations were so bizarre that at first we tended to disbelieve them.

Voyager 1 was launched from Cape Canaveral on September 5 1977. It was following in the wake of its twin, Voyager 2, which had begun its journey on the previous August 20; but Voyager 1 was moving in a more economical path. Both entered the asteroid belt between Mars and Jupiter on December 10; five days later Voyager 1 overtook its twin, and emerged safely from the asteroid zone on September 8 1978. In March 1979 it made its closest approach to the giant planet Jupiter, obtaining spectacular pictures as well as data; it then took advantage of Jupiter's tremendous gravitational pull to swing it out toward an encounter with Saturn. By November 1980 it had covered a track more than a thousand million miles in length.

Saturn is unique. It is smaller than Jupiter, but its equatorial diameter is almost 75,000 miles and its mass is 95 times greater than that of the Earth. Its surface is gaseous, with large amounts of the two lightest elements, hydrogen and helium; below the clouds comes a deep layer of liquid hydrogen, and there is a silicate core around 17,000 miles in diameter. The Saturnian 'year' is $29\frac{1}{2}$ times as long as ours, but the day is short and the rotation period is less than 11 hours.

The glory of Saturn lies in its ring-system. When the rings are suitably placed for observation a small telescope will show them, though when they are edgewise-on they almost vanish. The two bright rings, A and B, are separated by a gap known as Cassini's Division in honour of its discoverer, the 17th-century Italian astronomer G. D. Cassini. The inner ring, C, is more or less transparent. It has long been known that the rings cannot be solid or liquid sheets; a ring of this sort could not possibly exist so close to massive Saturn, and it had been found that the rings were made up of relatively small, icy particles. Pioneer 11, the first Saturn probe, passed by the planet in 1979 and confirmed the existence of two faint rings outside the main system; they were lettered E and F — D being reserved for yet another dim ring close to the planet, whose existence was dubious. Earth-based observers claimed to have recorded it, but I was sceptical, mainly because I had looked for it with very large telescopes under excellent conditions and had never been able to see it.

By November 6 1980 Voyager 1 was within six million miles of Saturn, and it

Saturn and its satellites Tethys (outer left), Enceladus (inner left) and Mimas (right of rings) are seen in this mosaic of images taken by NASA's Voyager 1 on October 30 1980 from a distance of 18 million kilometers (11 million miles).

Left *Details in Saturn's ring-system, as seen from Voyager.*

Right *Voyager view of the strange 'spokes' in Saturn's B ring.*

was already obvious that the ring system was much more complex than had been expected. And as the information continued to stream in to the Jet Propulsion Laboratory at Pasadena, California, all of us admitted that we were being taken completely by surprise. Instead of having only a few rings, Saturn proved to have many hundreds. To be more explicit, each of the known rings turned out to have many components, so that there were numerous ringlets and narrow gaps. I commented that the ring-system had more grooves than a gramophone record.

There were some observations which were particularly puzzling. First, there were the curious dark radial 'spokes' crossing the brightest ring, B. They persisted for hours after emerging from the planet's shadow, and when they disappeared they were replaced by new ones; but it was difficult to explain them, because ring-particles in the regions closer to Saturn move more quickly than those further out, and any 'spokes' should vanish quickly. Yet the features were there; and when Voyager passed 'below' the rings (that is to say, on the opposite side of the rings to the Sun) the spokes appeared bright instead of dark, so that presumably they were scattering the sunlight forward. My immediate instinct was to suppose that they were due to particles elevated away from the ring-plane by electrostatic or magnetic forces, and this does seem to be the best explanation. Saturn has a strong magnetic field, unusual inasmuch as the magnetic poles and the poles of rotation are almost in the same place.

Another shock was that two rings, one inside the C ring and the other in a dark gap at the outer edge of the Cassini Division, were eccentric rather than

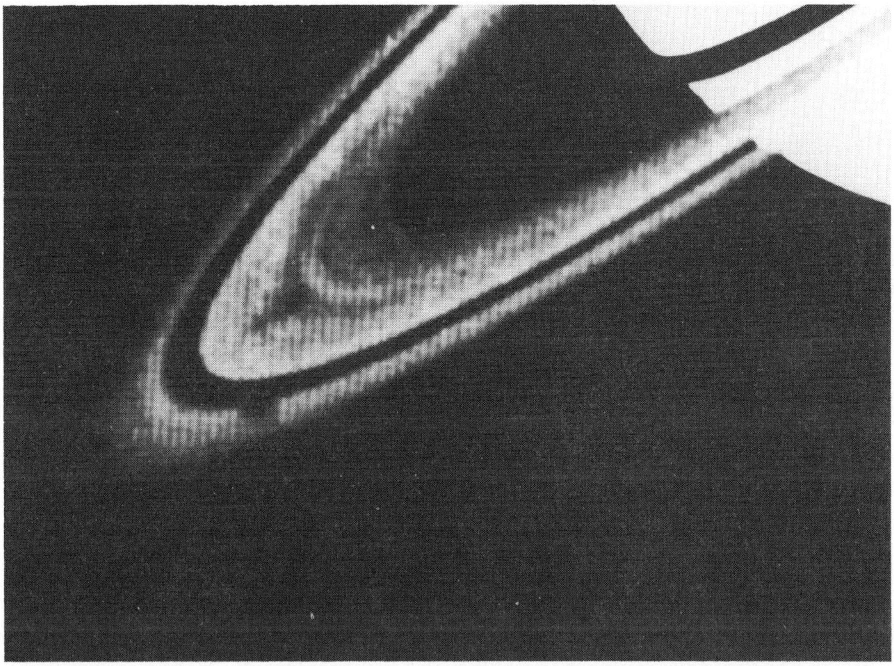

perfectly circular. The Cassini Division itself was not empty, but contained many thin ringlets. It had always been believed that the main divisions in the ring system, Cassini's, and Encke's Division in Ring A, were due to the perturbing effects of the inner satellites of Saturn, which would keep the divisions 'swept clear', but this explanation is certainly inadequate now. The satellites may be involved, producing a sort of wave effect in the ring system, but the full explanation is still not known.

Perhaps the oddest finding of all concerned the F Ring, which lies just outside the main system. It was known to be narrow; Voyager 1 showed it to be clumpy and braided, with three components, two of which looked as though they were intertwined like the strands of a rope. Dr Bradford Smith, head of the Voyager imaging team, commented that Ring F seemed to defy all the laws of celestial mechanics.

The revelations about the rings tended to divert attention from the globe of the planet itself, which appeared much blander than that of Jupiter, with more overlying haze. The temperature is much lower than that of Jupiter, so that clouds form deeper in the atmosphere and are partly hidden by small particles carried upward by convection. However, much detail was found. There was a distinct red spot in the southern hemisphere, several thousands of miles long, though much less striking and doubtless much less long-lived than the Great Red Spot on Jupiter. There were also interesting brownish ovals, as well as belts and bright zones. Like Jupiter, Saturn sends out more energy than it would do if it depended entirely upon what it receives from the Sun, but the cause seems to be different. Jupiter has not yet had enough time to lose all the heat which was built up during its formation; Saturn, smaller and less massive, has had ample time, and the excess energy is more probably due to the effects of droplets of helium sinking toward the centre of the planet through the lighter hydrogen.

Quite apart from the planet and its rings, there were the satellites. Nine were certainly known before the flight of Voyager 1; several more were discovered during the encounter, and the present total is at least 20. The system is quite unlike that of Jupiter. There is one very large satellite, Titan; then come Rhea and Iapetus (around 900 miles in diameter); Dione and Tethys (700 miles); and Enceladus and Mimas (300 miles), all the rest being much smaller. The

Mimas, from Voyager 1 — showing the huge crater Herschel, with its central peak. (The black spot is a fault in the picture.)

outermost satellite, Phœbe, was discovered as long ago as 1898, but is very faint indeed; it moves round Saturn in a retrograde direction, and may well be a captured asteroid.

Voyager 1 surveyed several of the satellites, with interesting results. Iapetus has one bright and one dark hemisphere; there are craters, and the surface is icy. Rhea, Dione, Tethys and Mimas are also icy and cratered; Mimas indeed has one huge crater (now named Herschel) which dominates the entire surface. Enceladus was not well seen from Voyager 1, but was quite clearly much smoother than the rest. Two small satellites were found to move at opposite boundaries of the curious F ring, and presumably control it, 'boxing in' the ring particles and pulling them back into the main zone if they show any tendency to escape. There were also two more small, irregularly-shaped satellites, now named Janus and Epimetheus, which have almost identical orbits. Periodically one satellite catches the other up, but there is no collision, and the two simply exchange paths. Very probably they are the halves of a former larger body which was broken up. Another satellite, not yet named, moves in the same orbit as Dione, and there are two Tethys 'co-orbitals', Telesto and Calypso. Telesto has a longest diameter of no more than 15 miles.

However, the main satellite target was Titan, which was already known to have an atmosphere. In Mission Control we were arguing about whether or not we would see the actual surface of the satellite; would the Titanian atmosphere be too cloud-laden? We soon realized that it was. All that could be made out was the top of a layer of orange 'smog'. But even here there were major surprises; the atmosphere proved to be thicker than that of the Earth, and to be made up chiefly of nitrogen, with a good deal of methane or marshgas. Everyone was glad that Voyager 1 had succeeded in sending back information about Titan. Had it failed to do so, then Voyager 2 would have to have been diverted past it, in which case it could not have gone on to rendezvous with the outer giant planets, Uranus and Neptune.

Voyager 1 had accomplished its task well, and it went on its way; no more planets were to be encountered, and the space-craft would leave the Solar System forever. Meanwhile Voyager 2 was following in its wake. It too obtained magnificent pictures of Jupiter, and then approached Saturn. The closest encounter was on August 25 1981, when Voyager was only 63,958 miles above the top of the cloud-layer.

As Voyager 2 neared its target, the images showed that the rings were even more complicated than had been thought. The radial spokes in Ring B were confirmed, with well-defined boundaries; there were thousands, not only hundreds, of ringlets and minor divisions. Improved views were obtained of several of the satellites, notably Iapetus, Tethys and Enceladus; Titan was not closely approached, but it was only too clear that Voyager 2 could have added little to what Voyager 1 had already told us about that orange-shrouded world.

Iapetus was of special interest. The difference between the bright hemisphere and the dark one was very marked; since the density is low (not much greater than that of water) it is safe to assume that the surface is icy, so that the dark material is a deposit of some kind. But how did the deposit get there? One idea was that it was wafted on to Iapetus from the surface of the more remote satellite Phœbe, but this seems unlikely, partly because Phœbe is so small and partly because it and Iapetus are so far apart (always at least six million miles). Alternatively, the dark material may have welled up from inside Iapetus, but there are difficulties here too, particularly since we do not know how thick the deposit is.

Hyperion, closer-in to Saturn, is smaller and is irregular in shape; it has been likened to a hamburger. Its longest diameter is not much over 200 miles. It is cratered, and one comment was that it looked like the broken-off section of some larger body. Unfortunately we are then faced with the problem of locating the other half, and there seems to be no sign of it.

Further pictures were obtained of the icy, cratered satellites Rhea, Dione and Tethys. Tethys has one huge crater, Odysseus, which is so large that if it had been formed by the impact of a meteorite the satellite would almost certainly have been broken up. The inner satellites Mimas and Enceladus are not alike. Voyager 1 had shown that Mimas, too, has one very large crater; Voyager 2 passed closer to Enceladus, which has only comparatively small craters and broad icy areas with no craters at all.

Up to the last part of the encounter all went well, but then there was a hitch; the scan platform, carrying Voyager 2's cameras, jammed. Valuable information was lost, but things might have been much worse, and as Voyager drew away from Saturn full contact was regained. As I write these words (mid-1984) Voyager 2 is en route for Uranus, which it will encounter in January 1986, and Neptune, which will be passed in August 1989. We must hope that all the instruments will continue functioning, despite that troublesome scan platform, because our information about Uranus and Neptune is very far from satisfactory, and no more probes to them have been funded as yet. But even if it does no more, Voyager 2, like its twin, has been an outstanding success and has done much to increase our understanding of the strange worlds in the outer part of the Solar System.

16 JODRELL BANK: END OF AN ERA

At the end of September 1981, Professor Sir Bernard Lovell retired as Director of the Nuffield Radio Astronomy Laboratories at Jodrell Bank — site of the world's most famous radio telescope. It marked the end of an era. Sir Bernard had been Director ever since the observatory had been established, and without his energy and skill the great 250 ft 'dish' would never have been built.

There are still some people who fondly imagine that one can look through a radio telescope. Of course one cannot. A radio telescope does not provide a visible picture of the object being studied, because its role is to collect radiations whose wavelength is too long to affect our eyes. In fact visible light makes up

250 ft Jodrell Bank Radio Telescope.

only a very small part of the total range of wavelengths, or 'electromagnetic spectrum'. To one side of the visible range we have infra-red, micro-waves and then radio waves; to the other side we have ultra-violet, X-rays, and finally the extremely short-wavelength gamma-rays.

Radio astronomy began half a century ago, more or less by accident. Karl Jansky, working for the Bell Telephone Laboratories in America, was investigating 'static', using an improvised aerial, when he discovered that he was picking up radio waves from the Milky Way. Strangely enough he never followed up his great discovery; he published a few papers, but they aroused little interest and Jansky turned his attention elsewhere. It was only during the war that the next major developments came. Radar was being developed (originally it was known as 'radiolocation') and it was found that the apparatus used for tracking German aircraft was being jammed. Deliberate interference was suspected, but before long it became clear that the interference was due to the Sun. One of the early investigators was Bernard Lovell, and after the end of the war he continued the investigations. Jodrell Bank was a good site; it was sufficiently far from Manchester where electric trams caused serious trouble. Preliminary equipment was set up, and the first results were encouraging.

By now Lovell was very much alive to the potentialities of radio astronomy, and after much thought he made a revolutionary proposal. He suggested building a fully-steerable radio telescope 250 ft in diameter. There was a good deal of scepticism, and several engineering firms said bluntly that the project was impossible. Finally Dr H. C. Husband, a well-known engineering expert, came to Jodrell Bank and was much less pessimistic. He described the problem as being no greater than that of putting a swingbridge over the Thames at Westminster. The original cost was estimated at a quarter of a million pounds, but before long, when construction work had actually begun, it was painfully clear that this had been a serious underestimate. Lovell was under no delusions; for a while he said nothing for the excellent reason that if he had not kept quiet, the whole project would have been stillborn. By the time that 'his iniquities were discovered', to use his own phrase, the telescope was reasonably well advanced.

There were endless crises. In particular, there was the problem of vibration. In America, the massive Tacoma Bridge had collapsed because of vibration due to quite low windspeeds; Husband built a model of the radio telescope, tested it in a wind-tunnel, and found that the danger was very real. Design modifications to overcome this unexpected difficulty cost £100,000. And as the total cost mounted the Public Accounts Committee became alarmed. There was an investigation. Due to a misunderstanding it was (quite wrongly) inferred that there had been inadequate liaison between Lovell and Husband, and at one time Lovell was faced with a personal bill for a third of a million pounds which he was certainly in no position to pay!

Then came an amazing stroke of luck. On October 4 1957 the Russians launched the first artificial satellite, Sputnik 1, and the barely-completed Jodrell Bank telescope obtained a magnificent trace of the carrier rocket; it was the only instrument in the world capable of such a feat. At once the Press, which had been antagonistic, changed its tune. Almost overnight Lovell was transformed from an eccentric spendthrift into a national hero. Money was still short, but finally, in 1960, Lord Nuffield came to the rescue. He sent Lovell a cheque for £50,000, enough to clear the deficit, and the real scientific work could begin.

Other artificial satellites and space-probes were tracked, but this was not the purpose for which the great 'dish' had been built. It was known that there were many radio sources in the sky, but their precise nature was uncertain. As time went by the situation became clearer, with Jodrell Bank very much in the forefront of research. The Sun was found to be a radio source, which was predictable enough, but most of the sources lay far beyond the Solar System. Some were associated with gas-clouds which were the remnants of supernova explosions, and the Crab Nebula was identified at radio wavelengths at an early stage. Others were associated with what seemed to be faint bluish stars, and in 1963 it was found that these objects were not stars at all; they were immensely luminous and immensely remote, shining more powerfully than any normal galaxies, and racing away from us at tremendous speeds. They were called 'quasars'. Today they are believed to be the nuclei of very active galaxies; they are certainly the most remote objects known to us. Later, in 1967, came the detection of 'pulsars', now known to be rapidly-spinning neutron stars. The discovery was made at Cambridge by Miss Jocelyn Bell (now Dr Jocelyn Bell-Burnell), but again Jodrell Bank was very much to the fore.

Radio astronomy has now become a vital part of research. Not all radio

Overleaf *The radio galaxy Centaurus A, once (wrongly) believed to be made up of two galaxies in collision.*

Below *Quasar 3C–273 — one of the quasars studied from Jodrell Bank at radio wavelengths. This optical picture was taken with the 200 in Palomar reflector.*

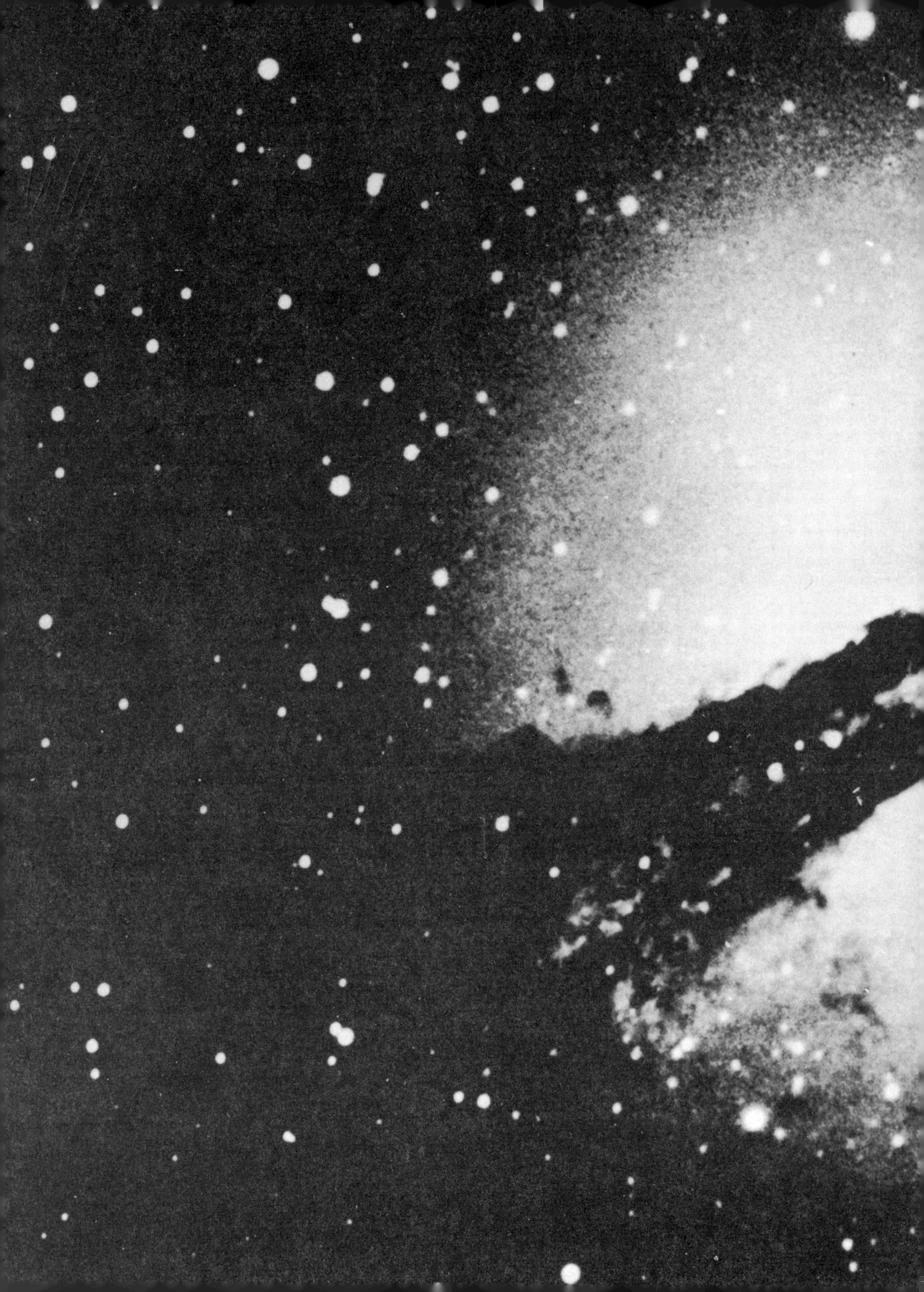

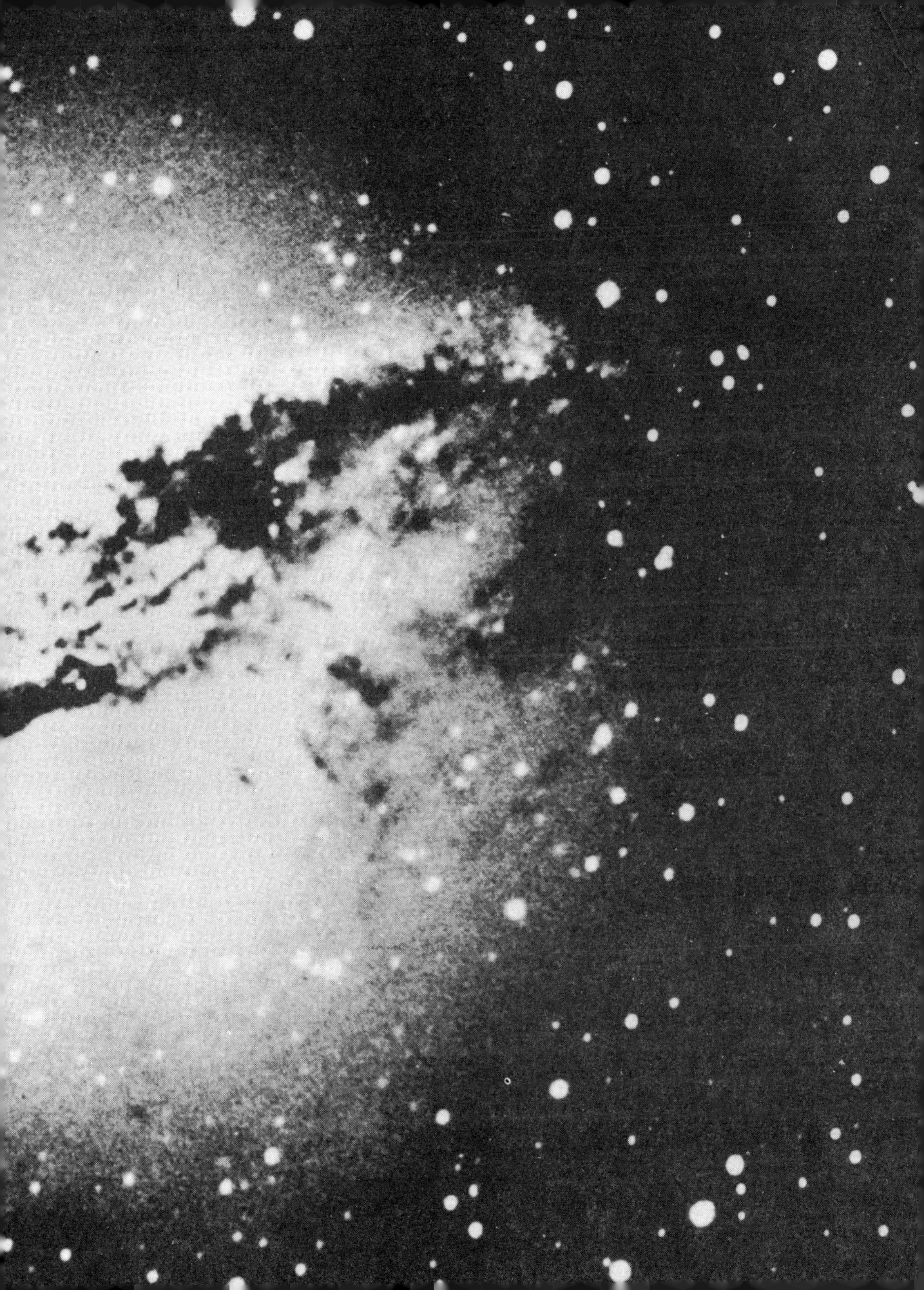

telescopes are of the 'dish' type, but all are used for essentially the same purpose. New equipment has been installed at Jodrell Bank, including a second large dish, and there is a link with other radio telescopes, which provides a tremendous increase in resolving power. Yet there may be trouble ahead. Civil and military radio interference is becoming more and more obtrusive, and Lovell has even voiced the fear that unless something is done, radio astronomy from Earth will become impossible within the next fifty years.

The only solution, then, is to take radio astronomy out into space — and plans are already being made, though as yet they are at an early stage. For instance, the far side of the Moon would be an ideal site, because it would be completely shielded from any Earth transmissions.

This lies in the future. Meanwhile, as Sir Bernard Lovell retired from his post as Director, he had every reason to be satisfied with his efforts. When he first put forward his idea for a great radio telescope, few people believed that he could succeed. But he did, and he opened up a whole new branch of research, without which modern scientists would feel very much at a loss. When the history of 20th century science comes to be written, Lovell will surely be remembered as the Isaac Newton of radio astronomy.

17 OPEN STAR-CLUSTERS

Look into a clear night sky during any evening from late autumn to spring, and you will be bound to notice a misty patch which is known as the Pleiades or Seven Sisters. It is conspicuous enough to attract one's attention, and a closer look shows that it is made up of stars. People with normal eyesight can see at least seven Pleiads under good conditions, and the record, set by a last-century German astronomer named Eduard Heis, is said to be 19. Binoculars reveal many more, and there can be no doubt that the Pleiades make up a genuine group, not a mere line-of-sight effect. We have here a splendid example of an open or loose galactic star-cluster.

Generally speaking, stars are widely separated in space. The Sun, with its attendant planets (including the Earth) lies well away from the centre of the Galaxy, in a region which may be described as of average stellar crowding; our nearest neighbour, Proxima Centauri, is over four light-years away (one light-year being equal to almost six million million miles). The stars in a cluster are considerably closer together than that, though they are in no danger of colliding with each other. In an open cluster, it is assumed that all the member stars were born out of the same cloud of dust and gas, and the Pleiades are comparatively young so that there is still a great deal of nebulosity spread between them. The principal stars are hot, white and energetic.

The brightest star in the Pleiades, Alcyone, is of the third magnitude, and is a fairly conspicuous naked-eye object. Next come Electra, Atlas, Merope, Maia and Taygete. Rather fainter are Celæno and Pleione, which are on the fringe of

The Pleiades or Seven Sisters.

naked-eye visibility. Pleione is of the special interest; it is what is termed a 'shell star', periodically throwing off rings of material, though these rings cannot be seen directly and have to be studied with instruments based upon the principle of the spectroscope. Pleione is also somewhat variable in light. Another of the named Pleiads, Asterope, is just about detectable with the naked eye by most people.

The nebulosity in the Pleiades is too dim to be seen with ordinary telescopes. The gas and dust is being lit up by the stars in and near it, making up what is termed a reflection nebula. Nebulosity of different type is found in the Sword of Orion, below the three stars of the Hunter's belt. Here we have extremely hot stars, which not only light up the nebulosity but also make it emit a certain amount of light on its own account, producing what is termed an emission nebula. But nebulæ of this kind are different from open clusters; they are stellar nurseries, where fresh stars are being born out of the interstellar material.

Another famous open cluster, well seen throughout evenings in winter, is that of the Hyades, extending from the brilliant orange-red star Aldebaran — easily found because the three stars of Orion's belt point directly toward it (in the other direction, toward the horizon, they show the way to Sirius in the Great Dog, the brightest star in the sky). The Hyades are much more scattered than the Pleiades, and make up a kind of V-formation extending to the right of Aldebaran. They are well seen in binoculars, but are somewhat overpowered by

The Hyades, extending to the right of the orange star Aldebasau near the centre of the picture: October 2 1971. The bright object to the upper right is Saturn. (Photograph by the late W. E. Pennell.)

Aldebaran. This is rather a pity, because Aldebaran is not a genuine member of the cluster at all; it merely happens to lie about half-way between the Hyades and ourselves, so that we are dealing with nothing more than a line-of-sight effect.

Also on view during winter evenings is Messier 35, an open cluster in the constellation of Gemini, the Twins; the name comes from the fact that it was the 35th entry in a famous catalogue of nebulæ and star-clusters drawn up by the French astronomer Charles Messier in 1781. M35 is just visible with the naked eye; telescopes and binoculars show it well. It is much further away than the Pleiades or the Hyades, at an estimated distance of 2850 light-years, and it contains well over 100 members.

However, all these open clusters are surpassed — in my view at least — by the pair in the Sword Handle of Perseus (not to be confused with the Sword of Orion), not far from the W of stars making up the constellation of Cassiopeia. The Perseus clusters were not included in Messier's list, but are just visible without optical aid on a clear night as a dim smudge. Telescopically, they show up as two separate clusters in the same low-power field of view; each is very rich, and they make a superb spectacle. Each cluster is some 70 light-years across; the distance from Earth is of the order of 7,000 light-years.

There are several other galactic clusters bright enough to be easily located. Præsepe in Cancer, the Crab, is one; it is often called the Beehive, and is at least 15 light-years in diameter. In condensation it is intermediate between the Pleiades and the Hyades, and it contains well over 50 stars. It has been known since very ancient times. Around the year 270 BC the Greek astronomer Theophrastus wrote that if Præsepe were not visible in a dark and apparently cloudless sky, rain was likely in the near future. There may be some truth in this; even very slight haze or high cloud is enough to make the cluster invisible without optical aid.

In the far south, never visible from England, we have the 'Jewel Box' in the Southern Cross, so called because among its bluish-white stars there is a prominent red one — Kappa Crucis. In the Scorpion we have two more lovely clusters, Messier 6 and Messier 7. Anyone equipped with a star-map and a pair of binoculars can find great pleasure in looking round the sky and seeking out other open clusters, some of which are beautiful to see and no two of which are alike.

Our Galaxy is a flattened system, with spiral arms, and it is in rotation. The Sun, at a distance of around 33,000 light-years from the centre, takes 225 million years to complete one circuit — a period often called the 'cosmic year'; one cosmic year ago the most advanced life-forms on Earth were amphibians, and even the celebrated dinosaurs had yet to make their entry. The age of the Sun is of the order of 5,000 million years, and stars in most open clusters, such as the Pleiades, are much younger than that. It seems that such a cluster is a temporary association. True, the stars in any cluster had a common origin, but the members of the cluster are not sufficiently close together to make up a stable system. The whole group will be constantly pulled upon by non-cluster stars

The double cluster in Perseus — the 'Sword Handle' (not to be confused with the Sword of Orion).

around it, and in the course of time the cluster will be so spread-out that it will lose its separate identity. Needless to say, this process is very slow and clusters well away from the more crowded parts of the Galaxy will persist for longest, but eventually dispersion must occur.

Today, unfortunately, good astronomical telescopes are expensive; one must be prepared to spend well over £100. However, binoculars are much cheaper, and are quite adequate to give really spectacular views. High magnification is not always an advantage; probably an ideal pair of binoculars for casual star-gazing is the 7 × 50 variety — that is to say a magnification of seven, with each main lens 50 mm in diameter. Armed with these, you will find many open clusters on view, together with double stars, coloured stars, and rich fields in and near the Milky Way. Should you acquire a pair, I recommend taking a careful look at the Pleiades, and seeing how many stars you can count. You will find a surprising number; the Seven Sisters have many junior relations!

18 MYSTERIOUS STAR

Capella, in the constellation of Auriga, the Charioteer, is one of the most brilliant stars in the sky. It is almost overhead as seen from England during winter evenings, and it never actually sets, though during summer evenings it skirts the northern horizon. It is yellow, like the Sun, but considerably more luminous. It is, incidentally, a close binary.

Close beside Capella lie three much fainter stars, known as the Hædi or 'Kids'. They make up a small triangle, and are very easy to see with the naked eye on a moonless night. One of them — Epsilon Aurigæ, at the apex of the triangle — is a most extraordinary system, and as yet we do not really know its true nature.

At a glance it looks normal enough. Generally it shines as a star of the third magnitude, rather brighter than the dimmest of the seven stars of the Plough; it is then slightly superior to Eta Aurigæ, at the left-hand base of the triangle, which is a normal star 370 light-years away from us and almost 600 times as powerful as the Sun. But Epsilon Aurigæ is not constant. Every 27 years it fades by about one magnitude, remaining at minimum for over a year before slowly regaining its lost lustre. Its strange behaviour was noticed by the German astronomer Fritsch as long ago as 1821, though it was not until our own century that much more was learned about it.

Epsilon Aurigæ is not a single star. The visible object is a slightly yellowish supergiant, around 60,000 times as powerful as the Sun and therefore the equal of Rigel in Orion. It looks inconspicuous because it is so remote. Its distance from us is of the order of 3,200 light-years, so that we see it today as it used to be about the time of the Trojan War. Associated with it is an invisible companion, and it is this companion which passes in front of the visible star every 27 years, cutting out part of the total light which we receive from the system. In fact Epsilon Aurigæ is an eclipsing binary, though of very unusual type indeed.

What exactly is the secondary component, which produces the eclipses? It is utterly invisible; it does not even radiate appreciably at infra-red wavelengths, but at least we can form some idea of its mass. With a binary system, the movements of the components relative to each other provide a clue to their masses; it seems that with Epsilon Aurigæ the visible star 'weighs' 35 times as much as the Sun, while the mass of the invisible secondary is 20 times that of the Sun. Few stars are as massive as this, and the secondary should certainly be detectable. Yet it is not. Therefore we are faced with a problem: what is dark, massive and invisible?

Originally it was thought that the secondary might be a very young star, still condensing out of interstellar material, and not yet hot enough to radiate. Its nuclear 'fires' would not have been triggered off. In this case the duration of the eclipses would tell us the diameter of the eclipsing body; it worked out at around 2,000 million miles, so that if it lay in the Solar System it would swallow up all the planets out to beyond Saturn! This would make it much the largest star known to science. Even the Red Supergiant Betelgeux, in Orion, has a diameter of less than 300 million miles.

At first sight this seems reasonable enough, but there are some disquieting facts to be taken into account. Even at mid-eclipse, half of the light of the bright component can still reach us, and it is difficult to believe in a large, invisible star which is also transparent. Moreover, it seems that no star as big as this could be stable, and we must cast around for another explanation.

The next possibility is that the invisible component of Epsilon Aurigæ is a Black Hole. Much has been heard of these bizarre objects in recent years (though there are several eminent astronomers who are decidedly sceptical about them). According to the generally-accepted theory, a Black Hole is produced by a very massive star which has exhausted all its sources of energy and has collapsed, becoming so small and so dense that its gravitational pull is tremendous; not even light can escape from it — and if light cannot do so, then certainly nothing else can, because light is the fastest thing in the universe. Therefore the old, collapsed star is surrounded by a kind of 'forbidden zone' from which nothing can escape.

A Black Hole itself may be small, but it can be surrounded by what is termed an accretion disk of dust and gas, and it has been suggested that the eclipses in the Epsilon Aurigæ system are caused by a disk of this kind. But again there are problems.

With a Black Hole, intensely heated material just outside the boundary would be expected to send out high-energy radiation, but nothing of the sort is observed from Epsilon Aurigæ. It is also hard to see how an accretion disk could remain uniformly cool and dark in the presence of so great an amount of gravitational energy.

The next theory is somewhat different. Instead of being very old, as a Black Hole would be, it is suggested that the invisible secondary is very young; it has not yet become a star in the accepted sense of the term, but is disk-shaped, so that it moves in such a way that it 'bisects' the visible star and allows some of its radiation to reach us. However, we would expect to find a rapid increase in temperature toward the centre of the agglomeration, and again nothing of the sort is found.

Astronomers are tending to revert to a theory which was first proposed 20 years ago, according to which the invisible secondary is a smallish, very hot blue star surrounded by a shell of gas. It is this shell, so strongly heated by the central star that it becomes opaque, which produces the eclipses. A hot star of this type would be expected to give out most of its energy not as visible light, but as ultra-violet radiation. The trouble is that ultra-violet radiation cannot be studied from ground level, because it is unable to penetrate the Earth's atmosphere, and the only solution is to observe from above the top of the air, which means using rockets or (preferably) artificial satellites. Fortunately this can now be done, and Margarita Hack, the Italian astronomer who first proposed the theory, has made use of the IUE or International Ultra-violet Explorer, a satellite launched specifically to carry out studies in the short-wave end of the spectrum. As yet the results are inconclusive, but there is enough evidence to show that the tell-tale radiation from Epsilon Aurigæ may exist, which lends support to Hack's theory even though we cannot be sure that the radiation comes from the mysterious secondary.

On the whole this explanation seems more plausible than the others, but it is not without its difficulties. Some hot blue stars are known to eject shells of gas, but all of those so far studied are less than one-thousandth as massive as the Epsilon Aurigæ secondary, and it is not easy to accept that a star can surround itself with a shell as substantial as this. We may learn more from the Space Telescope, the 94 in Hubble reflector due to be launched within the next few years.

The latest eclipse was that of the mid-1980s; starting on 22 July 1982, becoming total on 11 January 1983, ending its totality on 16 January 1984 and ending altogether on 25 June 1984. Henceforth nothing obvious will happen for another 27 years.

Young, embryo star? Black Hole, surrounded by an accretion disk? A disk of dust and gas which has not yet become a star? Or a hot star surrounded by a vast shell? The invisible member of the Epsilon Aurigæ system may be any one of these, or it may even be something quite different. At the moment we do not pretend to know, but at least we can be sure that Epsilon Aurigæ is unlike anything else that we have so far found in the universe.

19 WHEN PLANETS LOOP THE LOOP

During the early part of 1982 three bright planets — Mars, Jupiter and Saturn — were in the same part of the sky, in the regions of Virgo, the Virgin, and Libra, the Balance or Scales. All were very conspicuous, and almost everyone who looked up into the night sky will have noticed them. There was even an 'end-of-the-world' scare in some eastern countries, because it has been suggested that the pulls of the planets, acting in the same direction, might trigger off storms, cyclones and even earthquakes. Of course this idea was completely unfounded, but before saying more about the mythical 'planetary alignment' it may be helpful to discuss the movements of the planets in general.

The Earth is the third member of the Solar System in order of distance from the Sun. Mercury and Venus are closer-in than we are and therefore show phases from new to full, very similar to those of the Moon and for essentially the same reason; only half a planet can be sunlit, and the phase depends upon how much of the illuminated half is turned in our direction. The remaining planets lie beyond the Earth, and phase effects are negligible except in the case of Mars, which may sometimes appear the shape of the Moon a day or two from full.

With the exception of Pluto, which is probably not worthy of true planetary status, the orbits of the planets lie very much in the same plane. With reference to the orbit of the Earth, the angle of inclination is 7° for Mercury and less than

4° for all the rest, so that if we draw a plan of the Solar System upon a piece of flat paper we are not very far wrong. Obviously this means that the planets can be seen only in preferential directions. They keep to the Zodiac, a band stretching right round the sky. There are twelve Zodiacal constellations (actually a thirteenth, Ophiuchus the Serpent-bearer, does extend across the Zodiac between the Scorpion and the Archer) and the planets are always to be found somewhere in this band.

Our remote ancestors believed the Earth to be the centre of the universe; round it moved the Moon, the Sun, the planets and even the so-called fixed stars, which were assumed to be fastened on to an invisible crystal sphere. It was also assumed that all orbits must be perfectly circular — because the circle is the 'perfect' form, and nothing short of perfection could be allowed in the heavens! The change in outlook began with the work of the Polish cleric Copernicus in 1543, when he published his book claiming that the Sun must be the centre of the system of planets. Yet Copernicus was still wedded to the idea of circular orbits, and it was not until the early part of the following century that Johannes Kepler established that the planets move round the Sun in ellipses — though admittedly the orbital eccentricities are low.

In fact, the observed motions of the planets had always been difficult to explain upon the hypothesis of perfectly circular orbits and unchanging speeds. If this had been so, then the planets would have tracked steadily against the starry background, and this is not what happens. A superior planet (that is to say, a planet further from the Sun than we are) will seem to shift in an eastward direction against the stars, but will periodically stop and then slowly backtrack, moving in a westward or retrograde direction for some time before stopping once more and resuming its eastward movement.

Ptolemy, last of the great astronomers of the Greek school, was well aware of this, and tried to account for it by introducing numerous 'epicycles'. On the Ptolemaic theory, a planet moved in a small circular orbit around a centre or 'deferent' which itself moved round the Earth in a perfect circle. To account for all the observed phenomena, numerous epicycles had to be used, and the system became hopelessly clumsy and artificial, though it persisted for centuries after Ptolemy's death in or about AD 180.

The real facts are different. For instance, consider Jupiter's behaviour in 1982. At the start of the year it lay in Virgo; slowly it moved into the next Zodiacal constellation, Libra, but in February it reached a stationary point and then moved in a retrograde direction until it was back in Virgo. The next stationary point was reached in late June, after which Jupiter moved eastward once more into Libra.

Jupiter is much further from the Sun than we are; its mean distance from the Sun is 483 million miles, as against only 93 million miles for the Earth. Following Kepler's laws, Jupiter also moves much more slowly and takes $11\frac{3}{4}$ years to complete one circuit of the Sun. Therefore, some time before opposition, when the Sun, the Earth and Jupiter are more or less lined up, the Earth 'catches up' with Jupiter and passes it — so that Jupiter seems to loop and loop. The other superior planets behave in much the same way. Their revolution periods range from 687 days for Mars up to as much as 164¾ years for Neptune. (Pluto's period is 248 years, but this strange little world is now known to be smaller than the Moon, so that its status is very much in doubt.)

Now let us come to the alleged 'planetary alignment' which has become so notorious. In a book published some years ago entitled *The Jupiter Effect*, two science journalists, John Gribbin and Stephen Plagemann, claimed that in 1982 all the planets would be on the same side of the Sun and approximately aligned. This, they wrote, would affect the Sun, which would in turn affect the Earth, producing tremendous disturbances. They paid special attention to the San Andreas Fault, near Los Angeles and San Francisco, which is admittedly active; according to Gribbin and Plagemann the planetary alignment would very probably trigger off a disastrous earthquake there.

It is very difficult to see how this theory can ever have been taken seriously. The most elementary mathematics will show that all the planetary pulls combined are negligible compared with that of our nearby Moon, and could never produce the slightest disturbance on Earth. Apart from this, there was another slight objection: there was no lining-up of planets in 1982! The planets were spread out over an angle of more than 50°.

Unfortunately, once a scare of this sort has been started it is very difficult to stop. Many major observatories, including the Royal Greenwich Observatory, were bombarded with letters from people who were genuinely alarmed at the prospect. I must have had at least 500 letters myself, and the whole affair became an unmitigated nuisance. In several lesser developed countries there were even panics and demonstrations. Of course, it was not the first episode of its kind — there was another in 1917, for instance, caused by the writings of an amateur scientist named Porta — but this one was surprisingly widespread, presumably because the Gribbin-Plagemann book had been put out as a serious scientific forecast.

It is true enough that a major earthquake in California must take place sooner or later. But when it does you cannot blame the planets.

Though there was no lining-up, it was correct to say that during the late 1970s and early 1980s, the largest planets were on the same side of the Sun and spread out in a curve, which was very useful because it meant that a space-craft, Voyager 2, could be sent past Jupiter, Saturn, Uranus and Neptune in turn, using the gravitational pull of one to send the probe on to rendezvous with the next. The encounters with Jupiter and Saturn have already taken place. Uranus should be encountered in January 1989 and Neptune in August 1989, so that if all goes well we should be able to obtain close-range information from each of these lonely giants. But this is a very different thing from a dangerous alignment, which could never happen. Everyone will be heartily glad to hear the last of the absurd 'Jupiter effect'.

20 THE OBSERVATORY OF 'THE BOYS'

La Palma is not one of the most famous of the Canary Islands. Indeed most people confuse it with the town of Las Palmas. But today La Palma has a new distinction; it is the site of a great observatory, in which Britain is playing the leading part.

At the top of the main mountain are some rocks, known as Los Muchachos — the Boys. It is these rocks which have given their name to the observatory: El Observatorio del Roques de los Muchachos. (I am sure that there must be a local legend about 'the Boys', but nobody has been able to tell me what it is!) The height above sea level is 7,400 ft, and the scenery is remarkably picturesque. However, astronomers are not interested in scenery; what they want is good 'seeing' with clear, steady air, and the mountain-top certainly provides it.

The main role is being taken by the Royal Greenwich Observatory, which has itself had a long and varied history. It was founded in 1675, by express order of King Charles II, for navigational purposes, and is still the timekeeping centre of the world; all nations use GMT or Greenwich Mean Time, and one instrument at Greenwich, the Airy Transit Circle, is the zero for longitudes, so that one can straddle it with one foot in the eastern hemisphere and the other in the western.

Map of the island of La Palma.

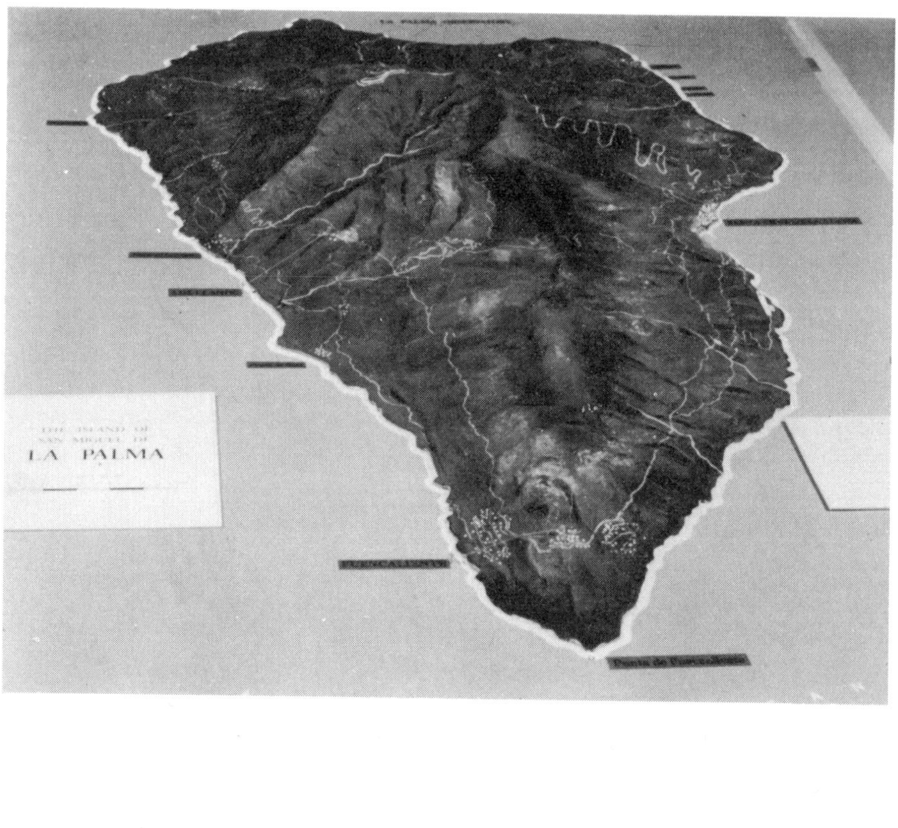

However, in our own century it became painfully evident that Greenwich Park was no longer suitable as a site for an observatory. There was too much smog and, above all, too much light pollution. The Old Royal Observatory was turned into a museum, and the actual working instruments were shifted to the clearer skies of Herstmonceux, near Hailsham in Sussex. It was here, in 1967, that the Isaac Newton Telescope or INT was set up. With its 98 in mirror, it was much the largest telescope ever erected in the British Isles.

Unfortunately, even Herstmonceux proved to be a comparatively poor site. There was too much cloud and too much glow, so that the INT could seldom be used to its full capacity, even though it was equipped with the latest electronic devices — which are rapidly superseding the sensitive plate for most branches of astronomy, just as the camera superseded the human eye a hundred years ago. Eventually the decision was made to search for something better, and the choice fell upon La Palma.

La Palma is a Spanish island, but the observatory has been organized upon a fully international basis, with Britain, Denmark, Sweden, Holland and other countries collaborating. There were many problems to be faced, even when the observatory was formally opened in May 1979. In particular, the road up to the mountain-top is decidedly rough; when I drove up it in late 1981 my Land-Rover became stuck on several occasions. Of course this is being dealt with, but it all takes time.

The INT was shifted from Herstmonceux to La Palma at an early stage, by which time the first telescope, a Swedish one, designed for studies of the Sun, was already operational. There were also plans for the great new William Herschel Telescope, which will be one of the largest in the world, and a Nordic telescope, a joint undertaking by Norway, Sweden, Denmark and Finland. All these telescopes are of the most modern variety. The astronomer of today does not have to sit in a cold dome for hour after hour; he is comfortably installed in a control room, and indeed he need not even be on the island at all. It will be possible to control the La Palma telescopes from Britain.

What kind of work will be carried out at La Palma? The main emphasis will be on stellar astronomy. Very large telescopes are seldom or never used for observations of our near neighbours, the Moon and planets; this would mean wasting their tremendous light-grasp. The principal aim is to study objects so far away that their light takes millions, hundreds of millions, or even thousands of millions of years to reach us. Our star-system or Galaxy is one of many. It contains about 100,000 million stars, of which our Sun is fairly typical, but we know of other galaxies which are far more populous. There are also the strange quasars, first identified in 1963, which are now thought to be the cores of very active galaxies, and are immensely remote and powerful. Recently, a quasar identified by astronomers in Australia has been estimated to lie at a distance of some 13,000 million light-years, so that it is well out toward the edge of the observable universe. It is racing away from us at about 95 per cent of the speed of light, so that it is much further away now than it was when you started reading this page.

Old INT dome at Herstmonceaux. The dome is now empty.

Many observatories are threatened by increasing light pollution. Even the great telescopes in California are not immune, because cities have an unpleasant habit of becoming not only larger (which does not greatly matter) but also brighter (which does). La Palma is in a favoured position. It is a small island, with no large towns anywhere near, and conditions there may be expected to remain good for a long time to come. The outlook is encouraging, and within the next few years we may confidently expect important scientific results from the 'Observatory of the Boys'.

21 THE HUNT FOR PLANET TEN

How many planets are there in the Sun's family? Most people will say 'nine', ranging from Mercury, much closer to the Sun than we are, out to remote Neptune and Pluto. Five of them are bright naked-eye objects, and have been known since the dawn of human history; they could hardly be overlooked. Three new planets have been found in more recent times: Uranus in 1781, Neptune in

1846 and Pluto in 1930. Today there is growing evidence that another planet exists, and for the first time there seems a definite chance that it may be tracked down.

Uranus was discovered by William Herschel more or less by accident (though it is only fair to add that he was carrying out a systematic review of the sky). Irregularities in the movements of Uranus led two mathematicians, Adams in England and Le Verrier in France, to calculate the position of a more distant planet, Neptune, which turned up very close to the position which they had given. Once more the Solar System was regarded as complete, but there were still small but definite irregularities in the motions of both Uranus and Neptune. In America, Percival Lowell made new calculations, and arrived at a possible position for a ninth planet. He had founded a major observatory at Flagstaff, in Arizona, and had equipped it with powerful telescopes, but despite all his efforts the new planet eluded him. He died in 1916; it was not until 1930 that Clyde Tombaugh, at the Lowell Observatory, discovered Pluto not far from the position given by Lowell.

Yet Pluto soon proved to be very much of a problem. It had an unusual orbit, with a period of 248 years, and at perihelion it was found to come within the orbit of Neptune. The next perihelion passage is due in 1989; between 1979 and 1999 Pluto temporarily forfeits its title of 'the outermost planet', though its

Dome of the telescope at Flagstaff with which Clyde Tombaugh discovered Pluto. (It is now at Anderson Mesa, away from the main observatory because conditions at Anderson are better.)

Above *Pluto and Charon, recorded in 1983 with a CCD on the Hawaiian telescope. The projections are spurious, but the two images are clearly separate.*

Below *Pluto, photographed with the Palomar 200 in reflector.*

orbit is tilted at the unusually sharp angle of 17° and there is no fear of a collision with Neptune. Even more perplexing was the revelation that Pluto is small. It is very faint (which is why Lowell himself failed to find it) and it is of low mass. In 1977 it was found to be attended by a satellite, Charon, and for the first time a reliable value for the mass and diameter could be given. Pluto cannot be more than 1,800 miles across, so that it is smaller than the Moon. Moreover, it is made up of a mixture of rock and ice, so that it is very much of a cosmical lightweight.

If so — and there now seems no doubt about it — then Pluto could not possibly exert any measurable pull upon giant planets such as Uranus and Neptune. In fact, it could not be the planet for which Lowell had been searching. Either the reasonably correct prediction was due to sheer luck, or else the real Planet Ten still awaits discovery.

We cannot rule out the first of these alternatives, but it would be an almost incredible coincidence, and I admit that I simply cannot believe it. Also, recent studies confirm that there really are unexplained perturbations in the movements of both Uranus and Neptune, in which case it is very likely that there must be an extra member of the Sun's family at a still greater distance.

The trouble is that to start hunting for Planet Ten without at least a vague idea of its position would be a hopeless business. It would occupy the whole time of a large telescope for years, with no guarantee of success. On the other hand, the irregularities in the motions of Uranus and Neptune are too slight to give us any reliable clue, and Pluto, which is probably unworthy of planetary status, can be ruled out.

Some years ago K. A. Brady, in America, studied the movements of Halley's Comet, which has a period of 76 years. At its greatest distance from the Sun it moves out well beyond Neptune, and Brady believed that he had identified slight irregularities which could be attributed to a new planet. However, it was later found that these calculations were unsound, and it does not look as though Halley's Comet can help us.

If natural bodies fail us, then what about artificial ones? This is an interesting possibility, and it so happens that there are two space-craft, Pioneers 10 and 11, which are ideal for the purpose.

Pioneer 10 was launched in 1972. In 1973 it by-passed Jupiter, sending back the first close-range information from that remarkable world, and then began a never-ending journey out of the Solar System; it will never come back.

Pioneer 11 followed a year later. It made its pass of Jupiter in December 1974, and was then given an extra burst of power which swung it back across the Solar System. It made a pass of Saturn in 1979, and it too has begun its final journey out of the Solar System. Therefore Pioneers 10 and 11 are leaving the Solar System in opposite directions. We are still in touch with them, and should remain so for some years yet — perhaps well into the 1990s.

Moreover, we know exactly how they are moving. This is measured by means of radio. A two-way radio signal is sent out at a known frequency; changes in frequency of the returning signal, due to the well-known Doppler effect, tell us

In April 1980 Pluto almost occulted a star. From my observatory in Selsey it was a disappointing 'miss', as shown in these drawings which I made with my 15 in reflector. Pluto is the smaller of the two images.

of any changes in the space-craft's velocity relative to the Earth. Plotted continuously, these velocity changes provide the exact trajectory relative to the known positions of the Sun and planets.

Suppose, then, that the Pioneers begin to wander? They could be used in the same way that Adams and Le Verrier used Uranus to track down Neptune, and they could give definite proof of the existence and even the approximate position of Planet Ten.

This was something that I suggested, in modified form, as long ago as 1974, but the fact that the two Pioneers are moving outward in almost diametrically opposite directions is sheer bonus. Suppose that one Pioneer is perturbed, and not the other? The cause could only be a planet moving beyond Neptune's orbit (bearing in mind that the mean distance between Neptune and the Sun is 2,793 million miles). True, the planet would have to be in the right place at the right time, but it might make its presence felt.

On the other hand, there is also the possibility that both Pioneers could show perturbations. To suppose that there would be two planets ideally placed is not reasonable, but there could be a much more massive body at a much greater distance. It could even be a dead star — a companion of the Sun which has run through its life-story and has lost all its energy, so that it is a dark, invisible Black Dwarf. In this case, its distance from the Sun would be of the order of 50,000 million miles.

A third idea is that the Pioneers might be affected by a very massive Black Hole at a distance of perhaps 100,000 million miles. Yet this seems much less plausible. A Black Hole is produced by a star which has collapsed, and is now pulling so strongly that not even light can escape from it; but we would expect at least traces of radiation from material about to be sucked into the Black Hole, and nothing of the kind has been observed.

Dead star, Black Hole or unknown planet? Each suggestion has its supporters. My own preference is for a planet, about the same size and mass as Uranus or Neptune, and moving round the Sun at a distance of around 5,000 million miles; but as yet we have too little to guide us. All we can say, with fair confidence, is that some unknown body exists in the far reaches of the Solar

System. If we are still in touch with the Pioneers in the mid-1990s, by which time they will have moved out well beyond Neptune, they may provide their final service to us by indicating the existence of Planet Ten — if, of course, Planet Ten is really there!

22 THE ASTRONOMERS OF NEW SOUTH WALES

Siding Spring Mountain, in New South Wales, is a peaceful place. The ancient volcanoes of the Warrumbungle range last roared out thirteen million years ago; today everything is quiet, and in the National Park kangaroos, wallabies and other kinds of roos are left blissfully undisturbed. Siding Spring itself is not really high; it rises to little over 4,000 ft, but it has achieved fame in recent years, because it is the site of what many astronomers regard as the world's best observatory.

Driving up the wooded road can be a somewhat hazardous business, particularly after dark, because kangaroos have remarkably little road sense, and are always apt to bound out without the slightest warning (which is why most cars have 'roo bars' fitted to their front bumpers). In the sunshine it is very attractive; gaily-coloured birds add to the scene, together with the occasional koala bear. Then, quite suddenly, you come in view of the large dome which houses the AAT or Anglo-Australian Telescope, the main instrument at Siding Spring.

The dome, 120 ft across, weighs a full 500 tons. The old volcanic rocks are completely stable, but to guard against any possible shaking the telescope is based separately from the actual building. In design the telescope is conventional enough. There is the usual skeleton tube and equatorial mounting; the telescope assembly weighs 335 tons. The heart of the telescope is, of course, the main mirror, which has a diameter of 153 in and is made of Cervit, which is much more suitable than old-fashioned glass. It has been said that the light-grasp of the AAT is so great that it could detect a candle-flame at a distance of 1,000 miles. It can certainly record stars which are ten million times fainter than those visible with the naked eye.

Astronomical techniques have changed beyond all recognition over the past few decades. At the end of the war, the observer had to sit in a cold, dark dome guiding his telescope while long-exposure photographs were being taken. Today all this has changed. Just as photography superseded the human eye toward the end of the last century, so photography itself is now being rapidly superseded by electronic devices, which are far more sensitive. The AAT itself is completely computer-controlled, and the observer spends his time in the adjacent control room, watching a television screen in comfort. To set the telescope, the required co-ordinates are plugged in; at the touch of a switch the

Dome of the AAT at Siding Spring.

telescope swings slowly but smoothly to the position indicated, and the dome turns until the slit is in the correct place. The setting is accurate to within a couple of minutes of arc, and a slight manual adjustment is enough to bring the target into the centre of the field of view. Then the driving mechanism takes over, and work can continue without any danger of the object drifting out of the field. Every possible precaution is taken to ensure complete accuracy. Computers allow for the very slight flexing of the mirror as the telescope moves around, and even changes of atmospheric pressure inside the dome are taken into account.

The telescope has been used to make some remarkable discoveries. It was here, for instance, that astronomers identified what is called the Vela pulsar. This is a stellar wreck — all that is left of a formerly very bright and massive star which blew itself to pieces in a supernova explosion, and was reduced to a tiny, rapidly-spinning, super-dense object made up of atomic particles known as neutrons. These pulsars emit radio waves, and hundreds of them have been found, but not many have been identified with visible objects. The first was in the celebrated Crab Nebula, the remnant of a supernova seen in the year 1054. The second is in the southern constellation of Vela. The radio position was known; the great power of the AAT allowed the object itself to be recorded. It is still the faintest visual object which has ever been tracked down.

Though electronic devices have come so much to the fore, 'old-fashioned' photography is not neglected at Siding Spring, where David Malin has used the AAT to take what are without doubt the most magnificent pictures of gaseous nebulæ and remote galaxies ever obtained. It is also helpful that the telescope is in the Earth's southern hemisphere, so that it can be used to study the stars of the far south which never rise above the horizon of the great northern observatories, such as those in California. We in England never see the Clouds of Magellan, the Southern Cross, the Centaur or the very richest parts of the Milky Way. We are also denied a view of Eta Carinæ, which is certainly one of the most fascinating objects in our Galaxy.

Eta Carinæ is about 6,000 light-years away, and is immersed in a mass of dust and gas; the so-called Keyhole Nebula. For a time during the last century it was the brightest star in the sky apart from Sirius, and was then estimated to be as powerful as about six million Suns put together. Since then it has faded, and is below naked-eye visibility, but its total energy output has not changed much and it is exceptionally strong in the infra-red. Recently there have been suggestions that it is approaching the climax of its brilliant career, and may be about to 'go supernova'. If so, it will shine for a brief period as brightly as a quarter-moon. The outburst may not happen for another century, another

The AAT.

thousand years or even a million years or so, but one can never tell. Visually Eta Carinæ looks quite unlike a normal star; I have described it as 'an orange blob'.

A few hours' drive from Siding Spring takes you to the little town of Parkes, where there is an obsevatory of a completely different kind. The main instrument is a 210 ft wire 'dish' used to collect radio waves from the sky. It came into operation in 1963, not long after the completion of the 250 ft radio telescope at Jodrell Bank in England, and has played a major role in all types of investigations. There is the closest possible collaboration with Siding Spring, notably with studies of pulsars and the remote, amazingly luminous objects known as quasars. The 210 ft telescope has been used to locate a pulsar in the Large Cloud of Magellan, a galaxy lying well outside the boundaries of our own at a distance of over 150,000 light-years. There has also been the case of the quasar known as PKS 2000-330.

Quasars lie at distances of millions or even thousands of millions of light-years. Even today we are by no means sure of their nature, though evidence is growing that they are the central parts of exceptionally active galaxies. Many of them are powerful radio sources. One of these was detected at Parkes in early 1982 and the information was passed straight to Siding Spring, where the AAT was used to track it down optically. Visually it looks like a dim, slightly blurred star. In fact it is believed to be about 13,000 million light-years away, making it the most remote object known to us. It is speeding away at about 95 per cent of the velocity of light, and cannot be so very far from the boundary of the observable universe.

The AAT is not the only telescope at Siding Spring. There are others, including the UK Schmidt, a special type of instrument operated by the University of Edinburgh, but the AAT reigns supreme. In sheer size it is easily surpassed by some other reflectors, notably the Russian 236 in in the Caucasus and the Palomar 200 in in California. But it is an open secret that the Soviet telescope is unsatisfactory, and the Palomar giant is now over 30 years old. All things considered, it is fair to say that even if the AAT has its equals, it has no superiors in the world today. It is a telescope of which its makers and users can be justly proud.

23 CELESTIAL CATHERINE-WHEELS

It has long been known that the Galaxy, of which our Sun is such an undistinguished member, is a flattened system. It is made up of about 100,000 million stars, some of which are a great deal more powerful than the Sun, together with clouds of dust and gas known as nebulæ. When we look along the main plane of the system we see many stars in almost the same line of sight, producing the lovely Milky Way band. The centre of the Galaxy is a somewhat mysterious place; we cannot see it directly, because there is too much obscuring material in the way, but we know that its distance from us is of the order of

The 'Whirlpool Galaxy' Messier 51 in Canes Venatici (the Hunting Dogs). It is 37 million light-years away, and is one of the finest of all the spirals. It was also the first spiral to be seen as such (by Lord Rosse, in 1845).

30,000 light-years. The entire Galaxy is rotating. The Sun takes approximately 225 million years to complete one journey round the galactic nucleus.

Seen from 'above' or 'below', the Galaxy would look like a vast spiral; the Sun lies near the edge of one of the spiral arms. Such a shape is not unexpected. Many other galaxies show spiral forms, sometimes face-on to us (as with the lovely

Left *The spiral galaxy Messier 101, in Ursa Major. It is over 11 million light-years away.*

Right *Spiral galaxies in Ursa Major — NGC 2976, 3031, 3034 and 3077. Taken with the Palomar 200 in reflector.*

Whirlpool Galaxy in the constellation of the Hunting Dogs) and sometimes placed at a narrow angle (as with the Andromeda Galaxy, which is decidedly larger than our own, and at its distance of 2,200 million light-years is one of the nearest of all external systems). Some of the spirals are 'tight-wound'; others are so 'loose' that the spiral arms are difficult to trace. There are also many galaxies which show no trace of spirality, and are either elliptical, globular or even completely irregular in shape.

Since our Galaxy is rotating, it would seem that the spiral arms should 'wind up' and disappear in a relatively short period, since the regions closer to the nucleus move faster than those further out (just as in the Solar System the innermost planet, Mercury, has the quickest orbital motion and the outer planets, Neptune and Pluto, the slowest). The age of the Galaxy is certainly well over 5,000 million years, and therefore we have to find some way of explaining the present Catherine-wheel shape.

The essential key must be linked with the movement of rotation. Elliptical and globular galaxies have almost no interstellar material left; all of it has been turned into stars, which is why these galaxies are often more luminous than the spirals. In the irregular galaxies, such as the southern Magellanic Clouds, there is still a good deal of interstellar matter remaining, but the systems are not rotating rapidly. Presumably these differences between types of galaxies are connected with their original mode of formation. We have to admit that the

basic causes are as yet poorly understood, but at least we are learning something about the evolution of the spirals.

What apparently happens is that as the system rotates, waves of density are set up in the gas, having some of the characteristics of sound waves. As the system rotates, the density wave pattern spins round at a different rate than the actual stars. Therefore, new stars form in these denser regions, and mark out the spiral arms. The arms begin to 'wind up', the very massive, luminous stars which have been produced in them soon use up their nuclear 'fuel' and disappear, either by exploding as supernovæ or else by losing their mass more gently in the form of what may be termed stellar wind. As the luminous stars disappear, the density wave pattern moves on, and the optimum conditions for star formation move along with it.

Consequently, no spiral arm is a permanent feature; but as old arms vanish as their highly luminous stars cease to shine, new spiral arms are formed. Triggering off of star formation may also be due to the death-throes of the previous generation, when supernova outbursts produce regions of compression.

All these ideas are comparatively new; the old theory — that elliptical galaxies evolve into spirals, or vice versa — has to all intents and purposes been given up, though it is true that some recent results from the Siding Spring Observatory in Australia indicate that in special cases there may be a complicated evolutionary story. All in all, however, we have to admit that our overall knowledge of the life-stories of galaxies is far from satisfactory and all branches of astronomy have to be used, from radio methods down to researches in the very short wavelengths. We have come a long way since 1845, when the third Earl of Rosse used his home-made 72 in reflecting telescope to see the first spiral forms in galaxies, but we still have a long way to go.

24 THE RETURN OF HALLEY'S COMET

On October 18 1982, three astronomers at Palomar Mountain, in California, made an exciting discovery. It did not look in the least spectacular; it was merely a tiny blur, at the very limit of the recording capability of the world's largest fully-operative telescope, the Hale 200 in reflector. But it was immensely significant. The dim blur was nothing more nor less than Halley's Comet, which had been out of sight for more than 70 years.

Comets have been termed the stray members of the Solar System. They are not solid, massive bodies like planets; a comet is made up of small particles, mainly ice, together with fine 'dust' and very tenuous gas. A major comet may have a long tail, stretching across the sky, though fainter ones may be entirely tailless, so that they look like small luminous patches. There are even some comets which seem superficially to be very like faint stars.

Comets shine by reflected sunlight, though it is true that when they are near perihelion (that is to say, the point of closest approach to the Sun) their gases give out a certain amount of self-luminosity. They are visible only when well inside the orbit of Uranus, which orbits the Sun at a distance of well over 1,700 million miles; and in general their paths are very elliptical.

Some comets take so long to complete one journey round the Sun that to all intents and purposes they may be regarded as non-periodical. We cannot predict them; they take us by surprise, and after a few weeks, months or years they pass out of view. However, there are also many comets with short periods, so that they return regularly. Encke's Comet, for instance, has a period of only 3.3 years and has been seen at more than 50 returns (in fact, it can now be followed all round its orbit). Crommelin's Comet has a period of 27 years; Olbers' Comet, 70 years — and so on. Unfortunately, with one notable exception all the periodical comets are dim objects, seldom or never visible with the naked eye. The one exception is, of course, Halley's Comet.

It has a long history. It was recorded by the Chinese several hundred years before the birth of Christ, and it has been observed at every return since that of 11 BC. The mean period is 76 years, though this is not quite constant and may vary by a couple of years either way. At some returns it may be magnificent, while at others it is comparatively inconspicuous. This is due not to changes in the comet itself, but to the relative positions of the comet and the Earth near the time of perihelion.

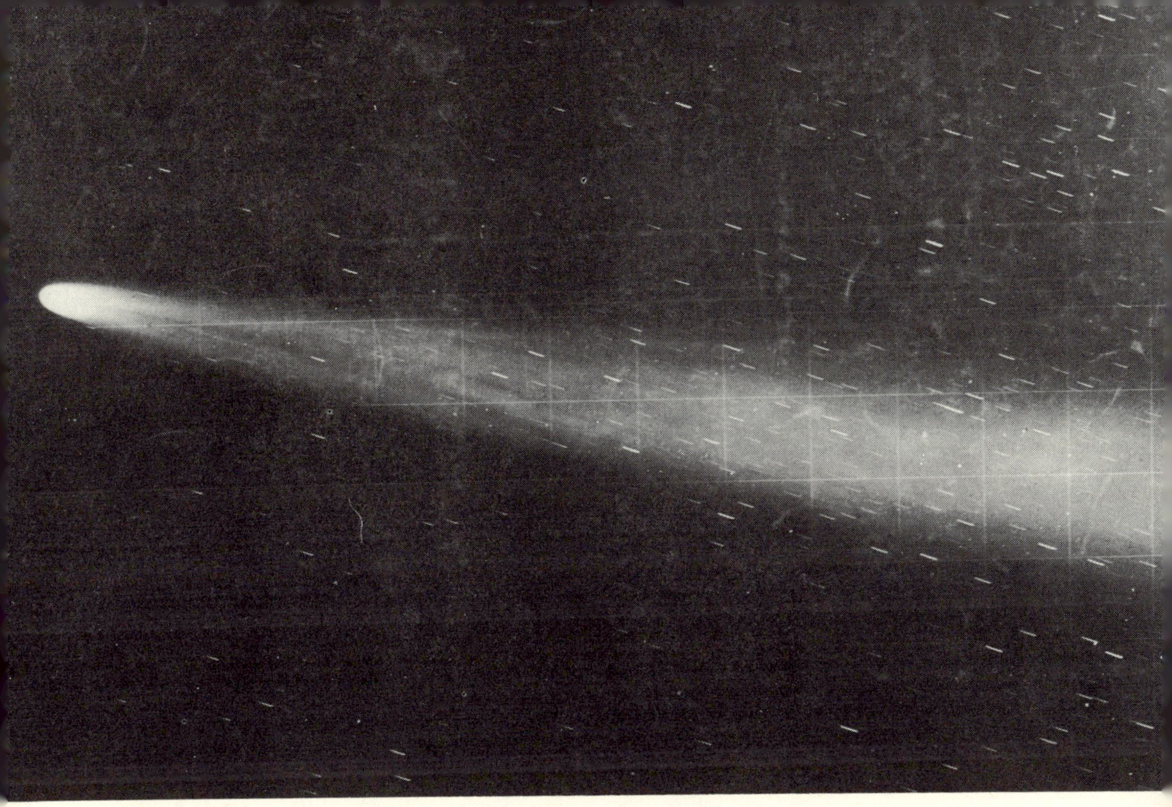

Halley's Comet, photographed from Mount Wilson in 1910.

At the return of 1682, the comet was carefully observed from England by John Flamsteed, the Astronomer Royal, and by Edmond Halley, who had already become well-known from his observations of the stars in the southern hemisphere of the sky. At that time it was not known that comets are members of the Solar System, and it was generally thought that they travel in straight lines. Halley was dubious. When he worked out the orbit of the 1682 comet, he found that it was practically the same as those of comets previously seen in 1607 and in 1531. Could they be one and the same? Finally he sent a paper to the Royal Society in which he wrote:

'Now many things lead me to believe that the comet of 1531, observed by Apian, is the same as that which in the year 1607 was described by Kepler and Longomontanus . . . I may, therefore, with confidence predict its return in the year 1758. If this prediction is fulfilled, there is no reason to doubt that other comets will return.'

Later he added, modestly, that if he were proved right, posterity would not refuse to acknowledge that the discovery had first been made by an Englishman.

Halley could not hope to live until 1758; he died in 1742. Subsequently some new calculations were made, and the date of perihelion was revised to 1759. On Christmas Night, 1758, a German amateur named Palitzsch picked up the comet more or less where it had been expected, and it duly passed through

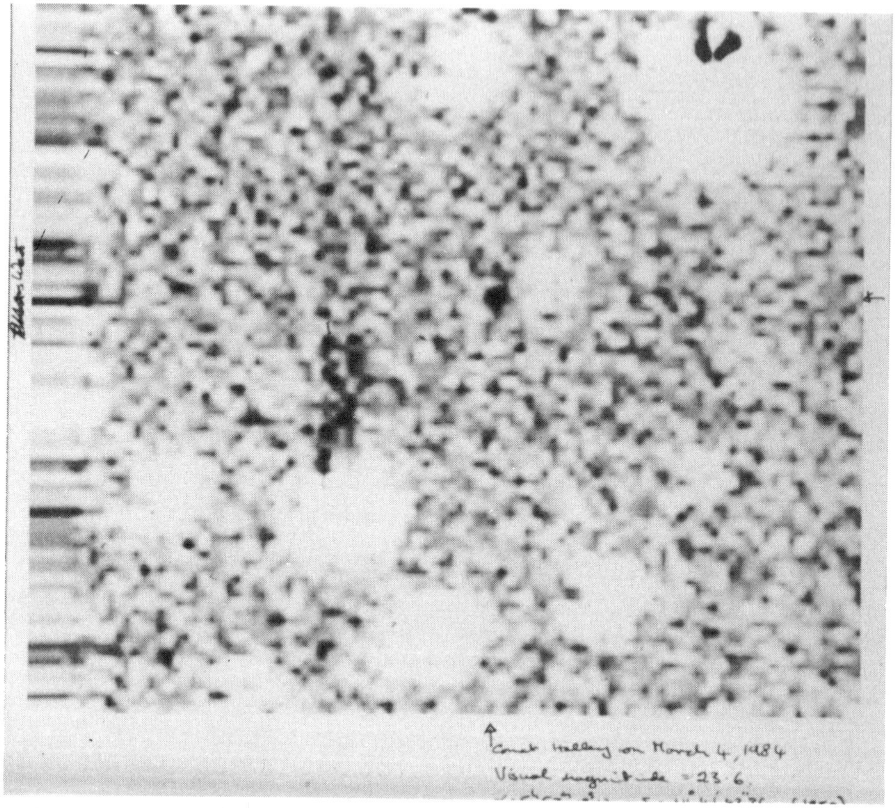

Halley's Comet in 1984. The stars are white blobs; the comet shows up as a dark object in the centre of the picture, which is of course a negative.

perihelion on the following 12 March. It was surely appropriate to refer to it henceforth as Halley's Comet.

Since then there have been two returns, those of 1835 and 1910. It was followed on the latter occasion until June 1911 — after which it was out of view until it was recorded again in October 1982 by the painstaking observers from Palomar.

We must admit, with reluctance, that the coming return is not a good one. The Earth and the comet will be in the wrong places at the wrong times. Halley's Comet will not be so striking as it has been on many previous occasions — as in 1066, when it shone down on Duke William preparing to launch his invasion of England, and was regarded by the Saxons as a warning of disasters ahead; or in 1456 when the current Pope, Calixtus III, preached against it as an agent of the Devil. It was also prominent in 1910, when the Earth actually passed through its tail and there was alarm in several countries including the United States. In fact, a comet's tail is so insubstantial that it can do no harm, and no effects whatsoever were recorded.

Yet the coming return will be of special importance, because of the rapid development of astronomical instruments and techniques since 1910. The Americans have abandoned their plans to send a space-probe to it (a decision which they will certainly regret for the next 76 years), but five missions will be dispatched, two by the Russians, two from Japan and the Giotto vehicle built by the European Space Agency.

The main task of these probes will be to investigate conditions inside the comet, and in particular to send back information about the nucleus — the very heart of the comet. At the moment we know very little about cometary nuclei because they are always surrounded by the dust and gas of the head, and are effectively hidden. We are not even sure of the size of a nucleus; it may be a few miles in diameter, but no reliable estimates have been possible. With luck, Halley will provide us with the answers.

It is not likely that Giotto, at least, will survive to emerge from the head of the comet, because it will penetrate much closer-in than the Russian or Japanese probes. A collision with a particle large enough to destroy the space-craft seems almost certain, all we can hope is that it will last for long enough to send back details about the hidden nucleus.

When the Palomar observers made their discovery, Halley's Comet was well over 1,000 million miles from the Sun — that is to say, between the orbits of Saturn and Uranus. Its magnitude was given as 24.3, making it one of the very faintest objects ever recorded. As it draws inward it will brighten, at first slowly and then much more rapidly. At its best, between November 1985 and the late spring of 1986, it will be a naked-eye object, apart from a period when it is on the far side of the Sun (perihelion is due on 9 February 1986). After that it will fade steadily, though powerful telescopes should keep it under observation until 1990 at least.

As Halley's Comet draws inward, and its tail starts to develop, it is safe to say that it will be studied more intensively than any comet has ever been. The fact that it was picked up so early means that it will remain in view for almost a decade. If we do not avail ourselves of the opportunity, we must wait until 2061 AD for another chance. But at least Halley's Comet itself will not fail us. The wanderer has returned.

25 SERPENT IN THE SKY

Whence come the comets? We have to admit that we are not sure; but most astronomers believe them to be genuine members of the Solar System, and that there is a cloud or reservoir of comets orbiting the Sun at a distance of a light-year or so (remembering that a light-year is equal to almost six million million

A modern 'meteor storm' — the Leonids, November 17 1966, seen from Arizona.

miles). When a comet is forced out of this 'Oort Cloud' — named after the Dutch astronomer Jan Oort, who first proposed it — it swings inward toward the Sun. It either makes one perihelion passage and then swings out back toward the Oort Cloud, or else it is violently perturbed by Jupiter or one of the other giant planets, so that it is forced into a short-period orbit.

When a comet becomes 'periodical', it naturally loses some of its material every time it passes through perihelion, because the icy materials in the nucleus evaporate to form a coma and a tail. This means that the short-period comets must be comparatively short-lived, though Halley's Comet, which has a period of 76 years, will presumably last for longer than most. Several comets which were seen regularly during the last century have now disappeared. Biela's Comet, with a period of 6¾ years, split in half during its 1846 return, and has never been seen since its next perihelion passage in 1852, though we can still trace its débris in the form of tiny shooting-star meteors.

Bennett's comet of 1970—one of the brightest in recent years.

Such is the general theory: the Oort Cloud has existed ever since the early days of the Solar System, at least 5,000 million years ago, and the periodical comets of today are ex-members of the Cloud which have been 'captured' by the planets. But a new, revolutionary theory proposed by two distinguished Edinburgh astronomers, Dr Victor Clube and Dr Bill Napier, takes a very different view.

According to Clube and Napier, the original Oort Cloud must have been dissipated long ago. Yet comets are still plentiful, and this means that the Cloud must have been replenished. Clube and Napier point out that there are times when our Sun, together with its planets, passes through a spiral arm of the Galaxy, where interstellar material is much denser than in the inter-arm regions. Each time this happens, some of the comets of the old Oort Cloud are lost — but many new ones, which had been moving between the stars, are picked up. For a few million years after the Sun's passage through a spiral arm, therefore, comets are particularly abundant.

For around 30 million years after the Sun has left an arm, the comet population is high, but as time passes the periodical comets lose their gas and dust, so that they become dead bodies. Can we see them? Clube and Napier believe so. Very small asteroids, a few miles in diameter, are known to pass near the Earth at times; they are known as Apollo asteroids, after the best-known member of the group. It is suggested that these are not true asteroids, similar to those in the main asteroid zone between the orbits of Mars and Jupiter, but are ex-comets. And because of their unusual orbits, they may occasionally collide with the Earth.

Such collisions unquestionably occur; for instance, the Arizona Meteor Crater was certainly produced by a gigantic impact about 22,000 years ago. Looking back further into the past Clube and Napier draw attention to an event which happened as long as 65 million years ago. The huge dinosaurs, which had dominated the world for so long, died out. It was not a gradual decline; it was a relatively sudden extinction, and it had already been proposed that this was due to a violent climatic change following the impact of an Apollo asteroid. There is indeed some confirmatory evidence. Bodies of this kind often include quantities of the rare element iridium, and the Earth's rocks at a level indicating that they were laid down 65 million years ago contain much more than their expected share of iridium.

We must also consider what are termed magnetic field reversals — when the compass needle changes from north-seeking (as at present) to south-seeking, or vice versa. These have certainly occurred, and Clube and Napier suggest that the cause is again an Apollo impact which, so to speak, churns up the iron-rich

The Aurora Borealis, April 13 1978, photographed from Scotland by K. Kennedy.

material deep inside the Earth and which is responsible for the overall magnetic field. One such reversal seems to have taken place at about the time of the dinosaur extinction, and another at the time of a much earlier extinction when many branches of living creatures vanished abruptly.

Finally, the two Edinburgh astronomers consider events in near-historic times. They believe that a very large comet was forced into an Earth-crossing orbit, so that it came regularly close to the Earth; it would then have been as bright as the full moon, with a brilliant tail stretching right across the sky, and it would have been accompanied by tremendous showers of meteors, far more splendid than the richest showers of modern times — such as the Leonids of 1799, 1833, 1866 and 1966. There would also have been many collisions with sizeable objects, though impacts from bodies as large as the 'dinosaur' period asteroid would have been extremely rare.

Clube and Napier believe that it was experiences at this time which led to the age-old fear of comets. The brilliant visitor — which they call the Cosmic Serpent — would have been accompanied by regular disasters caused by asteroidal bodies hitting the Earth. It is even suggested that the denser part of the meteor stream associated with the Serpent could dim the Sun for weeks at a time, and it is true that the ancient Greeks described eclipses lasting for a month or so — which no genuine eclipse can do. And even in 800 AD, Chinese astronomers were still looking out for high-level clouds which might dim the Sun.

Stone carvings dating from around 2000 BC are, say Clube and Napier, representations of comets, and the ancient Egyptians even portrayed the Zodiacal Light as a luminous pyramid. The Zodiacal Light is due to the Sun's rays being reflected from interplanetary material; today it is pale, but Clube and Napier suggest that for a thousand years or so, at the right seasons, viewers of the night sky would have seen two bright pillars of light stretching up along the ecliptic.

As the centuries passed by, the Cosmic Serpent lost its gas and dust; finally it changed into an Apollo asteroid, and Clube and Napier have identified it with Hephaistos, discovered in 1978, which moves in the right sort of orbit and is 6 miles in diameter. Its path is also very like those of Encke's periodical comet and the meteor shower known as the Beta Taurids. Many years ago F. L. Whipple, in America, showed that Encke's Comet and the Beta Taurids probably broke away from another body during the third millennium BC, and Hephaistos could have been that other body . . .

The whole theory is highly speculative; it is an attempt to weld astronomy, geology, palæontology, mythology, archæology and even history. Whether or not it will prove to be correct remains to be seen, but at least it is fascinating.

Can there be another Cosmic Serpent? According to Clube and Napier, this will almost certainly be the case next time the Sun passes through a spiral arm of the Galaxy and collects a new Oort Cloud. But that lies in the future, and there is no reason to fear that any major collision with an Apollo asteroid will occur in the near future. In our own age, there are no Cosmic Serpents in the sky.

26 THE LAST MAN ON THE MOON

In December 1972 Commander Eugene Cernan climbed back into *Challenger*, the Lunar Module of the Apollo 17 mission. Minutes later he and his companion, Dr Harrison ('Jack') Schmitt, blasted off to rejoin the third member of the team, Ronald Evans, who had been orbiting the Moon alone. The first phase in lunar exploration was over.

Gene Cernan was no stranger to space-flight. In 1969 he had been within a few miles of the Moon during the flight of Apollo 10 — the last rehearsal before the actual landing, achieved in July 1969 by Neil Armstrong and Edwin Aldrin. Moreover, Apollo 17 was scientifically the most valuable of all the missions, if only because Dr Schmitt was a professional geologist who had been given astronaut training specially for the occasion.

Ten years later I had a long talk with Gene Cernan, who is now a highly successful businessman. He has changed very little since his astronaut days, and, most significantly, he has lost none of his enthusiasm for the Moon. I asked him what was his most vivid memory, and I admit that his answer was not quite what I had expected. 'Overall,' he said thoughtfully, 'everything was rather overshadowed by the view of the Earth itself. It's been said that the Moon is colourless, bland and unbeautiful, but that's not true; the Moon is beautiful and

Patrick Moore and Commander Eugene Cernan, 1983.

majestic, with its valleys and its towering mountains. Yet to me, the greatest realization was that of standing in the middle of a dim charcoal picture, looking back through the blackness of space — blackness even though it's filled with sunlight — and seeing the Earth, with its life and its colour. That meant more to me than stepping out on to the surface of the Moon.'

One question often asked has been: Are stars visible in daylight from the lunar surface? Earlier astronauts had been non-committal, but Gene Cernan was more positive. 'When you're standing in sunlight on the surface of the Moon you can see stars if you concentrate very hard. Of course we made several orbits of the Moon before we came in for landing, and when you're over the dark side of the Moon you are probably in the blackest blackness anyone can imagine. You're out of sight of the Earth, and you can't even see the Moon below you. All you can see are hundreds and hundreds of stars.'

Obviously the landing area — known to astronomers as Taurus-Littrow, not far from the edge of the grey plain called the Mare Serenitatis or Sea of Serenity — had been carefully studied; Gene Cernan reckoned that he 'knew it as well as his own back yard'. Once on the surface, navigation had to be considered, particularly since the Lunar Module had carried a 'moon-car' in which Cernan and Schmitt were able to drive for considerable distances. Actually, estimating distances on the Moon is not easy, because there are no trees, roads or houses — and the Moon is a much smaller world than the Earth, so that its surface curves more sharply. 'When you look at some object and think it's about a kilometre away, it may really be as much as ten kilometres.' Of course the astronauts could make estimates by looking back at the Lunar Module, but there were times when they were behind huge rocks or ridges, and even the Module was lost to view.

The surface gravity on the Moon is one-sixth that of the Earth. 'That's much better than zero gravity, as in space,' Gene Cernan commented, 'because it's nice to know that there's an 'up' and a 'down', but there are problems. When you're walking, and want to change direction, you have to make up your mind several steps before you start to turn; when you're driving along in the Lunar Rover and hit a rock, one side of the Rover will rise up. You always have to be careful. The only thing between you and the vacuum of space is your space-suit, and to puncture that would be fatal.'

There was another problem, too — the lunar dust which he said 'was probably the most awkward hazard of all. The dust is like graphite; but graphite lubricates whereas lunar dust makes things stick together. It gets into your space-suits, and all moving parts of the vehicles. Back inside the Module, we took off our space-suits and we could actually handle the samples we'd collected (something which is never done by Earth-based scientists), and the dust is so fine that it even gets into the pores of your skin. It took me many weeks after my return to get rid of the last traces of it.'

During the Apollo 17 mission, I was at Mission Control in Houston, Texas, and I well remember the excitement when Dr Schmitt called out that he had found what he described as 'orange soil'. Nothing of the sort had been expected,

Earthrise over the Moon, as seen from the orbiting Apollo 17.

and it was confined to one area, the craterlet known unofficially as Shorty. 'When Jack yelled something about orange soil,' commented Gene Cernan, 'I reckoned he'd been on the Moon one day too long!' But it was unmistakable, and at first both the astronauts and the fascinated listeners in Houston believed that it might indicate recent volcanic activity, probably of the fumarole type. In fact this proved to be wrong. When the coloured material was analyzed, the orange hue was traced to numerous very small, very ancient glassy 'beads'.

As Gene Cernan pointed out, each Apollo mission extended the work of its predecessor. The first landing, with Apollo 11, was a there-and-back trip, designed to show only that the Moon could be reached. By the time of Apollo 14 (following the near-disaster of Apollo 13) the astronauts were able to equip themselves with a cart to carry their equipment around on the surface; Nos 15, 16 and 17 carried the Lunar Rovers, which were, of course, left behind and which will not deteriorate — so that one day, no doubt, someone will go up to them, fit them with new batteries and drive them off to a lunar museum. Apollo 17 was the climax of the entire programme. But when I asked Gene Cernan what he regarded as the most important results, he was decidedly cautious. 'I've often thought about that. I believe it may be many years before we really appreciate all the various aspects and decide which are the most significant. The true evaluation of our results may well be left to our grandchildren. I don't believe we're in a position to do it yet.'

Then, of course, comes the usual question: When will men go back to the Moon? Will it be relatively soon, or are we in for a long wait? Here again Gene Cernan was cautious, but not in any way pessimistic. 'I believe we'll go back. It may not be in my lifetime, or yours, because everything needs motivation. We went to the Moon not initially for scientific purposes, but for national and political ones — which was just as well, because it enabled us to get the job done! When there is real motivation, for instance to use the Moon as a base for exploring other worlds in the Solar System, or to set up a full-scale scientific base, then we'll go back. I can't tell when that motivation will come; it may be a long time, or it may not. I believe we'll go to Mars, too, but again there's the need for motivation. If the Viking landers had shown little green men with long ears looking back at us, we'd be on our way to Mars now. But eventually, yes we'll go back. There will be others who will follow in our steps.'

Would he like to return to the Moon? 'Yes, I'd go back tomorrow if I had the opportunity. I'd like to take time to recapture those moments which went by so quickly. Each of the 13 days of the mission had its own magic; but there were so many things which had to be done, and upon which your life depended, that

The Mare Serenitatis. This photograph, which I took with my 15 in reflector, includes the area where Apollo 17 landed.

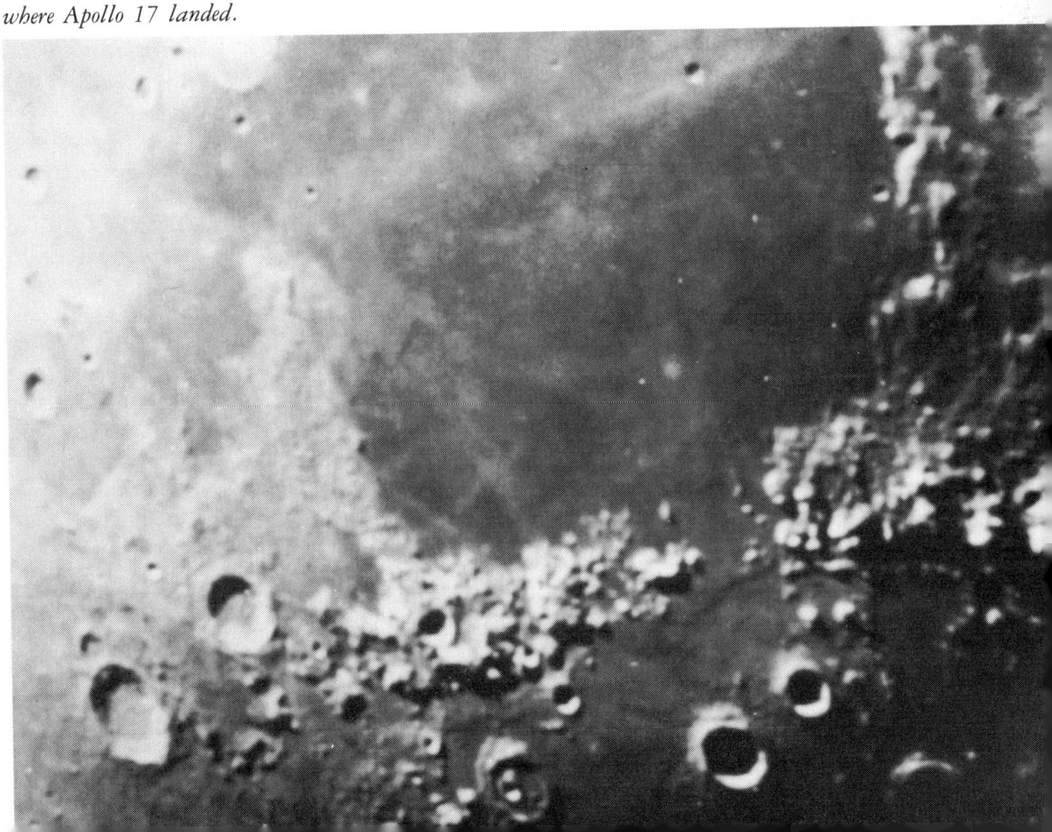

there was no chance to stand back and enjoy it all. I'd like to go back to do just that.'

To his regret, Gene Cernan knows that he will not make another flight to the Moon, but at least he and his companions have shown the way. Apollo has passed into history; for the moment, at least, Cape Canaveral lacks the urgency and excitement of those days when all thoughts were turned toward the Moon. But Apollo will never be forgotten. It marked the beginning of a new era in the story of mankind — and without men such as Commander Eugene Cernan, it would never have been possible.

27 RED SIRIUS?

Twinkle, twinkle, little star,
How I wonder what you are . . .

So runs the old rhyme. And stars do twinkle; there can be nobody who has not noticed it. Some stars twinkle more than others, but there is no star which shines with a completely steady light.

In fact, twinkling or scintillation has nothing to do with the stars themselves. It is due entirely to the fact that starlight has to come to us by way of the Earth's unsteady atmosphere, which, so to speak, 'shakes the light about' and causes the twinkling. From space, or from the surface of the airless Moon, there is no such effect.

It is easy to see that a star which is high in the sky twinkles much less than a star which is low down. The reason is straightforward enough. At low altitude, the star's light comes to us through a thick layer of the Earth's atmosphere, whereas from the zenith or overhead point the layer which has to be traversed is much less. During winter evenings, the yellow star Capella is almost overhead, while during summer evenings the position is occupied by the steely-blue Vega. Look at them at such times and the scintillation is not very noticeable, though it is still detectable. But when near the horizon, both Capella and Vega twinkle strongly. (They lie on opposite sides of the Pole Star, and about the same distance from it, so that when Capella is high up Vega is low down, and vice versa. Both are brilliant; indeed, the only stars which surpass them and are visible from England are Sirius and Arcturus.)

It is often said that a planet can be distinguished from a star because it does not twinkle. This is not entirely true. When a planet is low over the horizon, twinkling can be quite marked; but it is certainly less than with a star, because while a star appears as a point of light, a planet shows a small disk. The maximum apparent diameter of Jupiter, for instance, is 50 seconds of arc. This is not very much, but it does mean that the effects of the Earth's atmosphere are evened out to some extent.

Of all stars, the supreme 'twinkler' is Sirius in Canis Major, the Great Dog. It lies well south of the celestial equator, and therefore always appears rather low down from England. This, together with its exceptional brightness, causes violent scintillation, and Sirius seems to flash various colours of the rainbow, though it is really a pure white star.

There is something of a mystery here. Ptolemy, last of the great astronomers of Classical times, said that Sirius was red. Ptolemy (Claudius Ptolomæus), lived from 120 AD to 180 AD and was unquestionably a careful and accurate observer. Moreover, Lucius Seneca, about a century earlier, also said that Sirius was deep red. Much later, in 950 AD or thereabouts, the Arab astronomer Al-Sûfi drew up a catalogue of stars, and said nothing about Sirius' colour, so we may assume that he saw it as white as it is today. Either Ptolemy and Seneca were wrong, or else something very strange happened to Sirius between the years 180 and 950.

It would be easy to accuse Ptolemy of having defective eyesight — colour-blindness, for instance — but this explanation is completely out of court, because with all other stars his descriptions are correct. Betelgeux, Arcturus and Antares are given as red, while Rigel, Altair, Procyon and Regulus are white. Moreover, there are earlier references to a red Sirius to be found in Egyptian and Assyrian records.

We could perhaps find an explanation in the strong twinkling of Sirius. To go into a little more detail; the Earth's atmosphere acts rather like a prism, bending the various colours in increasing amounts from violet and blue at the short-wave end of the spectrum through to the red at the long-wave end; therefore, as the rays of starlight pass through the air, they suffer spreading-out or dispersion. This accounts for the multi-coloured flashing of Sirius, not nearly so noticeable with other stars because they are much fainter. Yet this does not seem to be the answer so far as Ptolemy is concerned, because he lived in Alexandria, where the latitude is only 31°N, and Sirius rises much higher in the sky than it can ever do from Britain — and the twinkling is correspondingly less.

Before jumping to the conclusion that Ptolemy was wrong, it is worth discussing the possibility that Sirius has really changed. It is a perfectly stable star, 26 times as luminous as the Sun and little over 8 light-years away, so that it is one of our nearest stellar neighbours; of the really bright stars, only the southern Alpha Centauri, at 4.3 light-years, is closer. (One light-year is equal to 5.8 million million miles, so that the distance of Sirius is approximately 50 million million miles.) There is absolutely no reason to suppose that an abrupt colour-change could have occurred, and frankly it seems to be absolutely out of the question.

However, Sirius is not a solitary traveller in space. It has a faint companion with only 1/10,000 the brightness of Sirius itself; it was discovered in 1862 by the American optician Clark, who was testing a powerful new telescope which he had made. (Remember, though, that the existence of the Companion had been predicted earlier because of slight irregularities in the individual or proper motion of the bright primary.) Since Sirius is often called the Dog-Star, the Companion has been nicknamed the Pup. At first it was assumed to be cool and red, but then, in 1915, spectroscopic observations by W. S. Adams in the

United States showed that its surface is hotter than that of the Sun. The Pup is white. And if it is hot and faint, it must be small, with a diameter of only about 26,000 miles — less than that of a planet such as Uranus or Neptune. Yet its mass is the same as that of the Sun, and the density must be enormous. It works out at 60,000 times that of water, so that several tons of it could be packed into an eggcup.

Initially, astronomers were sceptical; but eventually it became clear that the density really is of such an order. According to modern theory, a star of mass comparable with that of the Sun condenses out of interstellar material, shrinks and becomes hot. When the central temperature has reached about ten million degrees, nuclear reactions are triggered off. Stars contain a great deal of hydrogen, the lightest and most plentiful of all the elements, and in a star's core the hydrogen is being changed into helium, with release of energy and loss of mass. When the supply of available hydrogen is used up, the star changes its structure. The outer layers expand and cool; the interior contracts and heats up, so that different reactions begin. The star has become a Red Giant, as Betelgeux is today. At last, after a long and complicated series of nuclear reactions, what is left of the star becomes a very small, super-dense 'White Dwarf'. All the parts of the atoms are broken up and packed together with almost no waste space, which explains the amazingly high density. The Sun will become a White Dwarf one day, though not, fortunately, for at least 5,000 million years in the future and probably longer.

Now let us look back at the system of Sirius. The Pup is a White Dwarf, so that it must have passed through its Red Giant stage. Can this explain the colour as described by Ptolemy?

It sounds an attractive theory — but it simply does not stand up to analysis. For one thing, the combination of the present Sirius with a Red Giant companion would mean that the total light would be much greater than that of the planet Venus, and the star would be visible in daylight; the old astronomers would certainly have made comments. Secondly, the time-scale is all wrong. Though the transition from Red Giant to White Dwarf may be quick by cosmical standards, it would take thousands of years, if not millions, certainly not only a few centuries. Neither can we believe that the Pup suffered an explosion of the type we call a nova. Again the brightness would be too great.

Some years ago, during a *Sky at Night* programme, I asked viewers to write in and tell me what they made of the colour of Sirius (I carefully avoided explaining just why I wanted to know). I had over a thousand replies. Everyone noted the strong twinkling, but basically the answer was 'white' and there was no case of any observer who described Sirius as red.

All in all, then, I think we must resign ourselves to the fact that there has been no change in historical times, and that Ptolemy was either wrong or else has been misinterpreted (his great book has come down to us only by way of its Arab translation; the original Greek has been lost). There is no scientific explanation of how Sirius could have changed from red to white over so short a period. All the same, it is an intriguing puzzle!

28 IRAS

The early morning of January 26 1983 was an important one astronomically. At 2.17 am IRAS, the Infra-Red Astronomical Satellite, was launched from the Vandenberg Air Force Base in California. The take-off was perfect, and before long IRAS was settled in its orbit, moving round the Earth at a height of approximately 560 miles in a period of 103 minutes.

Exact timing was essential; the so-called 'launch window' lasted no more than 12 minutes, because IRAS had to be put into a very particular kind of path, carrying it over each pole and therefore staying close to the demarcation line between the daylit and night hemispheres of the Earth. The advantage of this so-called 'sun-synchronous' orbit is that the satellite is lit for most of the time, and can draw power from its solar panels, though very careful precautions have to be taken to prevent the delicate equipment from becoming overheated.

The IRAS satellite — the latest vehicle of this kind, and much the most sophisticated to date — is a combined American-Dutch-British experiment. The vehicle is controlled from the Rutherford Appleton Laboratory at Chilton in Oxfordshire. The information collected during each orbit is stored on tape, and played back each time IRAS passes over the Chilton area. The altitude is also carefully worked out. If IRAS moved at a lower level, there would be trouble

Comet IRAS–Araki–Alcock of 1983: the first comet discovered by IRAS.

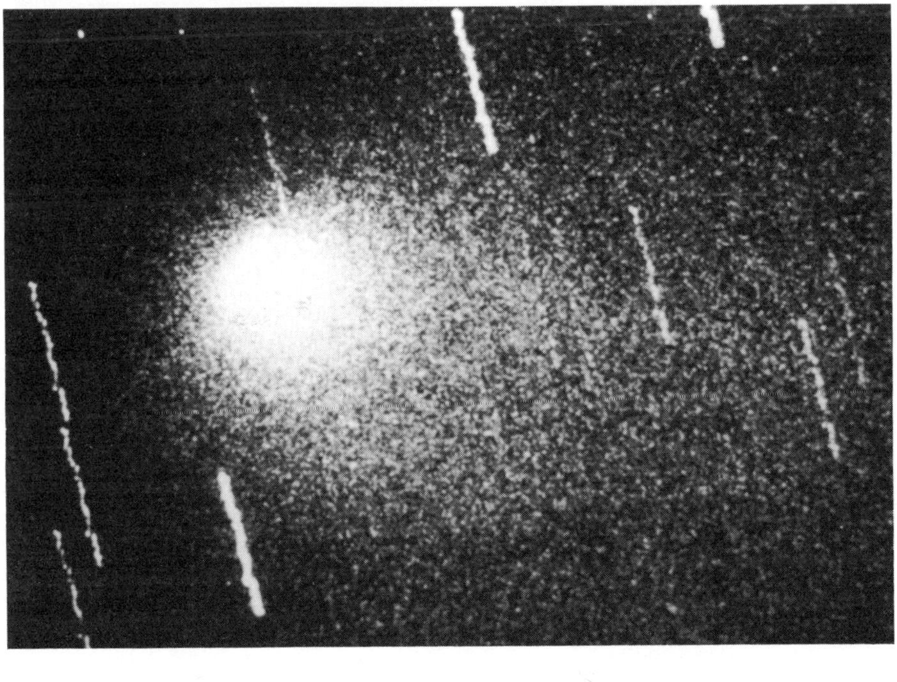

from the upper part of the Earth's atmosphere; but if it ventured too high, it would be affected by the zones of radiation which are known as the Van Allen zones in honour of Dr James Van Allen, the scientist who was in charge of the programme which led to their detection from the first successful American satellite, Explorer 1, more than two decades ago now. Without claiming that the 560 mile orbit is ideal, it seems to be the best possible compromise.

Unfortunately, IRAS has a limited active lifetime. This is bound up with the very nature of its main instrument, a 22.4 in infra-red telescope. Basically, an infra-red telescope looks rather like an ordinary optical telescope. The radiation is collected by a curved mirror and then directed on to a second mirror, which sends it back to the detectors through a hole in the main mirror. Of course it does not produce a visible picture, because infra-red radiation does not affect the eye. It may be called 'heat', with a wavelength longer than that of visible light but shorter than that of radio radiation.

By these standards, a telescope at ordinary room temperature is effectively white-hot, and would completely mask the much weaker infra-red radiation coming from the sky. Consequently, there has to be a very efficient cooling system. Liquid helium is used; this has a temperature of -270°C, only 2° above absolute zero (that is to say, the lowest temperature possible) which is enough to lower the temperature of the equipment to about 10° above absolute zero. However, the liquid helium inevitably boils away, which means that the lifetime of a telescope such as that in IRAS must be short.

Some branches of infra-red astronomy can be carried out from the Earth's surface, but even a trace of atmosphere means that some of the most important wavelengths are blocked out, whereas IRAS is unhampered by any such problem. Its main task is to carry out an infra-red survey of the entire sky. It is amazingly sensitive; it has been said that it could detect the radiation from a football at room temperature over a distance equal to that between London and New York!

There are, of course, infra-red sources in the Solar System, but the main work is concerned with objects much further away. For instance, there are infra-red sources deep inside the Great Nebula in Orion, quite invisible optically because they are hidden by the material of the nebula itself. A nebula of this sort is termed an HII region, because the hydrogen atoms in it are broken up ('ionized') by the radiation from very hot stars — in the case of the Orion Nebula, by the stars which make up the famous Trapezium, easily seen with a small telescope because they lie on the Earth-facing edge of the nebula. But there are many other HII regions; one of these, Sagittarius B-2 in the constellation of the Archer, is thought to have a total mass three million times that of the Sun concentrated in an area less than 20 light-years in diameter.

There are also vast clouds of molecules in the Milky Way. These molecules range from hydrogen to water, ammonia, carbon monoxide and even organic molecules such as formaldehyde. They are extremely cold (often no more than 10° above absolute zero) and may well be in the early stages of collapse, so that eventually stars will form in them.

IRAS, the Infra-Red Astronomical Satellite.

Some Red Giant stars have dusty shells; the most famous example is Betelgeux in Orion — some 15,000 times as luminous as the Sun, and large enough to contain the entire orbit of the Earth. Then we find stars of different type, which have come to a late stage in their evolution and have shed thick layers of dust which are opaque enough to block out all the direct light from the old star itself; however, the starlight is converted to infra-red as it reaches the surrounding dust. There are also very young stars, which too are surrounded by dusty shells but are in the process of blowing them away; and there are dark globules, irregular in outline, which seem to be proto-stars. Optically, they betray themselves only because they blot out the light of stars beyond, but they can be studied in infra-red. Surveys will be made of groups of galaxies, many millions of light-years away, to see whether any 'proto-galaxies' can be found, not yet detectable in visible light. There is no definite proof that these exist, but by now astronomers have learned to expect the unexpected!

Until little more than half a century ago, astronomers were limited to studying the universe in visible light. From the early 1930s radio astronomy was developed; today, thanks to the new techniques, we can study almost the whole range of wavelengths, from the ultra-short gamma-rays through to the very long radio waves. Space research methods have played an invaluable role, and the results from IRAS are leading to new, highly important advances.

29 HALF-WAY TO SPACE

Most people associate the islands of Hawaii with blue skies, guitars, hula-hula girls and sunny beaches — the ideal place in which to spend a holiday. Of course this is true enough, but there is another part of Hawaii's Big Island which tourists seldom see from close range — and probably do not want to do so. It is a region of cold, strong winds which spring up with amazing rapidity, and thin air. Here, on the top of Mauna Kea, astronomers have set up one of the world's leading observatories.

Mauna Kea is a typical shield volcano, 13,800 ft high. It has not erupted for so long that it may be regarded as extinct, though its companion volcano, Mauna Loa, is still active; once during the 19th century, lava from Mauna Loa even menaced the outskirts of Hilo, the only sizeable town on Big Island (and, according to local legend, was halted only by the timely intervention of a powerful witch-doctor who was hastily summoned to the scene). Kilauea, on the flanks of Mauna Loa, is always liable to erupt. However, Mauna Kea is pleasingly calm. Its total height from the sea-bed to the summit is 32,000 ft, so that as a structure it is loftier and more massive than Mount Everest.

Astronomically, the advantage of observing from the summit is that at least 40 per cent of the atmosphere lies below and, more importantly, the crest is

Domes on Mauna Kea.

Dome of the UKIRT on Mauna Kea.

above 90 per cent of the water vapour contained in the Earth's air. Water vapour is the enemy of the astronomer who wants to study the infra-red radiations coming from space. Infra-red is absorbed by water vapour, so that there is really no alternative but to rise above it. IRAS and other satellites have been used, but obviously they are limited with regard to the equipment they can carry. Mauna Kea is not restricted and in 1964 Gerard Kuiper, one of the world's most far-sighted astronomers, realized that it would be an ideal vantage point.

There were many problems to be faced. Inaccessibility was one; a road had to be built and even now the last part of it is very rough. All supplies and equipment had to be brought up from Hilo, and sudden storms can be dangerous; on one occasion three astronomers were marooned on the summit for several days (fortunately after the main domes had been built). But the main hazard is the altitude. The air is thin, and one's intake of oxygen is much less than normal. Different people react in different ways, but great care is needed, and to go to the summit and then start running up and down ladders is emphatically not to be recommended. Moreover, nobody sleeps in the observatory itself. The living quarters are at Hale Pohaku, just below 9,000 ft above sea-level, which is comparatively safe. The usual procedure is to spend at least 24 hours in Hale Pohaku before driving the rest of the way to the summit,

My photograph of the UKIRT on Mauna Kea.

which, rather surprisingly, takes less than half an hour. I was unaffected; others were not so lucky.

There are four main telescopes at the present time. Pride of place goes to the UKIRT — the United Kingdom Infra-Red Telescope, which has a mirror 150 in in diameter, and is the largest telescope of its kind in the world. Next comes the CFH or Canada–France– Hawaii 144 in reflector. NASA operates a 120 in infra-red telescope, and finally there is the 88 in reflector of the University of Hawaii. Seen from a distance, the view of these great domes strung along the spectacular and often snowy volcano slopes is spectacular by any standards.

UKIRT, funded by the Science and Engineering Research Council and operated by the Royal Observatory in Edinburgh, is of unusual construction. It is a lightweight telescope, with a much thinner mirror and a less substantial mounting than for a conventional instrument of the same size. It was designed to study infra-red radiation, whose wavelengths are much longer than those of visible light; therefore, the mirror need not be so accurate. Most people are familiar with infra-red (after all, one can feel the heat from the bars of an electric fire after they have been switched on well before they start to glow). The UKIRT is a Cassegrain; that is to say, the light (or infra-red) passes down an open tube, strikes the 150 in mirror, is reflected back to a second mirror in the tube itself, and is then passed into the detectors via a hole in the main mirror.

In fact, UKIRT has proved to be so good, and so accurate, that it can be used for optical work as well as infra-red — an unexpected bonus, casting great credit upon the skill of its makers.

As with all infra-red equipment, the detectors have to be kept at a very low temperature, only a few degrees above absolute zero. Also, the infra-red emissions coming from the sky background must be eliminated. This is done by tilting the secondary mirror to and fro. In one position the detectors record the object under study; in the other position only the sky is 'seen', and the infra-red due to the emitter can be measured by means of subtraction.

Though UKIRT weighs 80 tons in all, it has to be moved with absolute precision. Occasionally there are ground tremors, and eruptions from Mauna Loa have been known to produce inconvenient dust and ash clouds, but on the whole there has been no real trouble — yet.

The CFH 88 in is a conventional telescope, but it too can be used for infra-red, and some interesting results have been obtained recently by Dr Dale Cruikshank, who has confirmed suspicions that the surface of the strange little planet Pluto is covered with a layer of methane frost. It was also with the CFH that two French astronomers, M Bonneau and M Foy, obtained pictures showing that Pluto really does have a satellite, Charon. Like Pluto, Charon is probably made up of a mixture of rock and ice.

The telescopes on Mauna Kea were quick to prove their worth. As an observing site, the top of the volcano is as good as any location in the world, and from the point of view of the infra-red astronomer it is unrivalled. Therefore, it is natural that plans are already being made to set up new telescopes of various kinds not on the actual summit — most of the 'room at the top' has already been taken — but slightly lower down, in what is already being called Submillimetre Valley because the telescopes there will operate at wavelengths of rather less than a millimetre. After all, it has been said that to an astronomer the top of the volcano, at almost 14,000 ft above sea-level, is half-way to space!

30 THE QUICKEST PULSAR

In 1967 a remarkable discovery was made by a group of radio astonomers at Cambridge. They were using specialized equipment for a more or less routine investigation when they discovered a strange radio source which seemed to be 'ticking'. Its pulses were so rapid and so regular that for a brief period it was even thought that they might be artificial. However, this intriguing idea was soon abandoned. What the Cambridge group had found was a pulsar and before long others were detected; by now the total has reached well over 300.

We have also found that pulsars are the remnants of former massive, luminous stars which have exploded as supernovæ, and literally blown themselves to pieces. The Rosetta stone to the pulsars was provided by the Crab Nebula, in the constellation of Taurus. Today it consists of a mass of gas, spreading outward from its centre; it is certainly the remnant of a supernova which was observed by Chinese astronomers in the year 1054 — though since the Crab is 6,000 light-years away, the actual outburst took place before there were any skilled star-gazers to observe it. And at its centre there is a pulsar.

A supernova outburst occurs when a massive star uses up all its available nuclear 'fuel'. There is a sudden collapse, followed by a shock-wave which disrupts the star. At the core, the various atomic particles are crushed together to form neutrons, which have no electrical charge, so that the actual remnant of the supernova is a neutron star. Because there is little waste of space, the density of neutron-star material is quite remarkable. A cupful of it would weigh thousands of millions of tons.

The pulsars have very powerful magnetic fields; 'beams' of radiation are emitted in two opposite directions, from the magnetic poles, and since the pulsar is spinning quickly we have what may be termed a searchlight effect. Every time a 'beam' sweeps across the Earth, we receive a pulse of radiation. In the Crab Nebula there is a pulsar with a spin-rate of 30 revolutions per second. It has also been identified with a very faint, flashing object, and in fact only one other normal pulsar has been optically identified; this is in the southern constellation of Vela. The others are known only because of their radio signals.

It soon became evident that all pulsars are slowing down, and therefore losing their energy. Of course the rate of slowing-down is very slight, but it can be measured, and it follows that over a sufficiently long period a pulsar will lose all its remaining energy. Until recently the Crab pulsar held the record for the fastest rotation. This also showed that pulsars must be small — only a few miles in diameter; a larger body spinning at such a rate would become unstable.

Just as astronomers were beginning to feel confident that they understood these strange neutron stars, a new discovery was made; that of the pulsar known by its catalogue number of 1937+21, not far from the famous coloured double star Albireo in Cygnus, the Swan. The pulses were found to be incredibly rapid, 642 per second, so that clearly the object was in a different category from all other known pulsars. At its estimated distance of 8,000 light-years it was quite invisible optically, and its discovery was rather in the nature of an accident, but already it has presented theorists with many problems.

There is no doubt that it is a pulsar, and its diameter cannot be more than 20 miles or so. It has been carefully monitored at Jodrell Bank, and apparently the slowing-down rate is very slight — much less than for any other pulsar. Obviously, the rate of slowing-down is a key to the pulsar's age, and thanks to the Crab Nebula we have one definite example from which we can calibrate others. For 1937+21, we find an age of at least 100 million years. This seems very improbable, and so it is generally thought that the pulsar must have been speeded-up in some manner.

One theory suggests that the neutron star is a member of a binary or two-star system, the other component being a rarefied Red Giant. The immensely strong gravitational pull of the neutron star pulls material away from its companion; this material forms a cloud round the neutron star, and since both components are in motion around their common centre of gravity the cloud rotates, imparting extra spin to the neutron star itself. Alternatively, we may be dealing with a binary system made up of two neutron stars which affect each other. Yet a third possibility is that two neutron stars have actually collided.

From the way in which the object 'twinkles' at radio wavelengths, it has been possible to measure its speed through space; approximately 70 kilometres or 43 miles per second, which is quite normal for a pulsar. But in almost all other respects 1937+21 is in a class of its own. There seems every reason to suppose that its magnetic field is very weak by pulsar standards, even though it must still be about 100 million times as powerful as the magnetic field of the Earth. Presumably it is the remnant of a supernova outburst, but even here we cannot be sure, because not all supernovæ produce pulsars. For instance, the last two supernovæ to be seen in the Galaxy were those of 1572 (observed by the great Danish astronomer Tycho Brahe) and 1604 (seen by Johannes Kepler). In each case the site is now marked by a radio source, but not by a pulsar.

It is fair to say that the discovery of 1937+21 was unexpected. There had been no serious thought that a pulsar with such rapid rotation might exist, and it was sheer luck that 1937+21 lay close beside another radio source of more normal type. So long as we have only the one example to guide us, our knowledge must remain very incomplete, and therefore serious efforts are now being made in England, Holland and elsewhere to locate more pulsars of the same kind. Much of the sky, particularly the Milky Way region, will be surveyed. There is absolutely no reason to suppose that 1937+21 is unique, and it is very probable that many more quick-spinners will be found in the near future.

As yet 1937+21 has been under study for only a short while, but it may be assumed that its behaviour at the present time is typical of it. Since it is slowing down so very gradually it will continue sending out pulses for many centuries or even thousands of centuries, but eventually it must lose all its remaining energy, and will end up as a tiny, massive, inert body, completely undetectable because it will send out no radiations at all. If this is so, then there may well be many other dead pulsars moving through space. We may be sure that there are no such objects in the neighbourhood of the Solar System, because they would make their presence felt by their gravitational pulls, but sooner or later it is possible that we will encounter one of them.

Until more of the new-type pulsars are found there is little more to be said, but at least 1937+21 has opened up new avenues of research. Before long we may hope to find out more about these strange, dying stars.

31 THE WORLD'S STRANGEST OBSERVATORY

Most of the world's major observatories are built on the slopes or the summits of mountains. The Mauna Kea Observatory in Hawaii has an altitude of almost 14,000 ft, which can pose problems for some of the astronomers who go to work there. At the other end of the scale there is a remarkable observatory in South Dakota which is actually below ground. It lies at a depth of 4,850 ft — and yet it is from here that some of the most important observations of the Sun are being made.

South Dakota has a romantic history. Little more than a century ago it was the country of the gold-miners and also the gunslingers; Calamity Jane, Wild Bill Hickok, 'Doc' Holliday and the rest. Deadwood Gulch was the centre of their operations, and Wild Bill was actually killed in a gunfight in a Deadwood saloon on August 2 1876. Today the gunslingers are no more than a memory, but the gold is still there, and Homestake Mine, opened in 1877, is in full production. It is in fact the largest mine in the United States; it is situated between Deadwood and the little town of Lead (pronounced 'Leed'). The solar observatory has been set up in a pair of chambers hollowed out at the end of one of the passages. This level has been more or less worked out, and the present mining is carried on at a greater depth.

But why should students of the Sun have any wish to burrow a mile underground, where there is no hope of seeing the sky at all? To explain the reason for this apparently eccentric behaviour, we must look at the way in which the Sun — a typical star — produces its energy. It is not 'burning' in the conventional sense of the term, even though its surface temperature is almost 6,000°C and the temperature at the core rises to well over 10 million degrees. It may better be described as a vast nuclear reactor. It contains a great deal of hydrogen, the lightest and most plentiful substance in the universe, and near the core very strange things are happening; atoms of hydrogen are banding together to make atoms of the second lightest element, helium. (To be more accurate, we should say hydrogen nuclei, because at such temperatures the complicated atomic structures are broken up.) Each time a new helium nucleus is formed from four hydrogen nuclei, a little energy is released and a little mass is lost. It is this energy which keeps the Sun radiating, while the mass-loss amounts to four million tons per second. The Sun 'weighs' much less now than it did when you started to read this page, though fortunately it is so massive that there will be little obvious change for at least 5,000 million years in the future.

On the other hand, we cannot pretend that we have solved all the Sun's secrets. Far from it! In particular, the energy-production should be accompanied by the emission of neutrinos, which are particles with no electrical charge and virtually no mass. This makes them extremely difficult to detect, since they can pass right through the Sun and also through the Earth without being checked. However, there is one way to catch them. A neutrino may occasionally interact with the nucleus of an atom of chlorine, and in this case the

Outside Homestake Mine, South Dakota.

chlorine atom will be changed into a different kind of atom, argon-37, which is radioactive and is therefore comparatively easy to track.

To detect solar neutrinos, a great deal of chlorine is needed. The best way to provide this is to use a large tank filled with perchloroethylene, which is nothing more nor less than cleaning fluid. Its chemical formula is C_2Cl_4, showing that each molecule is made up of four chlorine atoms together with two atoms of carbon. Therefore, if sufficient quantities of perchloroethylene are used, it should be possible to see how many chlorine atoms are hit by neutrinos and changed into radioactive argon.

This is all very well, but the experiment cannot be carried out from ground level, because particles known as cosmic rays, coming from space, produce the same kinds of effects. The only answer is to provide a screen against the cosmic rays — and hence the underground observatory in Homestake Mine. Few cosmic rays can penetrate a mile of solid rock, whereas neutrinos can do so with no trouble at all.

The pioneer work was undertaken by Dr Raymond Davis and his colleagues at the Brookhaven National Laboratory in the United States. During the 1950s the plans were drawn up; the authorities at Homestake Mine were most co-operative, and in 1964 the underground observatory began its career, using a tank containing no less than 100,000 gallons of cleaning fluid. It is this which may be called the observatory's 'telescope', certainly the most unusual in the world. Obviously the whole procedure is delicate and difficult. The tank is left for a period of about two months, and then tested to see how many atoms of argon-37 have been produced. The whole tank holds about a thousand million million million atoms of various types, and what has to be done is to detect about ten atoms of argon-37 which have been created by the incoming solar neutrinos.

When ready for testing, the fluid in the tank is mixed by pumping helium into it, and the gases are fed through into the adjacent control room, where they are caught in a vessel which contains charcoal. The argon-37 sticks to the charcoal, and when the charcoal is subsequently heated the argon is given off, to pass through various other refining devices before entering a container. This container is taken out, and put into a lead box thick enough to shield it from all other radiations. The counts of the radioactive argon-37 atoms are then checked on a ticker-tape.

The results have been startling. There is sharp disagreement between theory and observation; the Sun is sending out only about a quarter as many neutrinos

In Homestake Mine — showing the huge tank of cleaning fluid used to trap solar neutrinos.

as it ought to do. Either there is something wrong with the experiment, or else there is something wrong with our theories about the mechanism of the Sun.

The results at Homestake have been absolutely consistent, and have been confirmed by a similar experiment carried out in the USSR. So many checks have been made that it is very difficult to believe that the experiment can be faulty. Therefore, we must look hard at our theories about how the Sun behaves.

It has been generally accepted that the temperature at the core of the Sun is about 15,500,000°C. If we reduce this to 14,500,000° the neutrino results fall neatly into place, but a reduction of a million degrees would raise all manner of other theoretical problems. Alternatively, we may be making errors in our estimates of how material in the Sun's globe is being mixed. It is even possible that neutrinos break up during their 93 million mile journey from the Sun to the Earth, but this is something which physicists as a class are reluctant to accept. Finally, there is a chance that the Sun is really behaving in an abnormal way at the present time.

We know that in some respects the Sun is a variable star. It reaches maximum activity every eleven years or so, when sunspots are frequent; the last maximum occurred in 1980, so that the next may be expected about 1991. However, there is excellent evidence that for a long period, between 1645 and 1715, this semi-regular cycle was suspended, and there were almost no sunspots at all. Earlier spotless periods have also been suspected, though with less certainty because the records are so incomplete. In fact, we can no longer be confident that the solar cycle is permanent, in which case many long-established theories may have to be drastically revised.

Perhaps the greatest importance of this work lies in the fact that neutrinos come to us directly from deep inside the Sun, in which respect they are unique. Until we can solve the problem which Homestake Mine has raised, we are very much in the dark.

We on Earth depend entirely upon the Sun. Comparatively slight variations in its output could have disastrous results — and it is possible that such variations have been responsible for the Ice Ages which have affected the Earth now and then throughout its history; the last Ice Age ended a mere 10,000 years ago. Therefore it is essential for us to find out as much as we can about the way in which the Sun radiates, and the apparent paucity of neutrinos is extremely puzzling. It is easy to see why astronomers are paying so much attention to a tank of cleaning fluid buried deep below the Black Hills in South Dakota. To us, the underground observatory is more important than the gunslingers ever were.

32 SPECKLES OF LIGHT

It must be said that to an astronomer, the Earth's mantle of air is an unmitigated nuisance. Not only does it make images unsteady, but it also blocks out many of the radiations coming from space. Going above the atmosphere is one solution, but even from the Earth's surface new techniques have come to the fore. One of the most interesting of these is known as speckle interferometry.

Astronomical photography is limited in scope because of the atmosphere. An exposure-time of as much as one second results in blurring of the image, and conventional photography cannot provide a resolution of less than about one second of arc. This is a small quantity — it is about the same as the angular diameter of a 2p piece seen from a distance of three miles — but it is too large to satisfy the astronomer. For example, no star shows an apparent diameter of as much as one second of arc.

An ordinary photograph of a star shows a symmetrical blob, often with protruding spikes. The spikes, of course, are due to the equipment used; and the larger the blob, the brighter the star — it has nothing whatsoever to do with the star's angular diameter. It is generally true to say that the smaller a star looks, either when photographed or when observed directly, the better you are seeing it. And if the exposure of the photograph is longer than about 1/50 of a second of time, the unsteadiness of the atmosphere produces a smeared image.

With a very short exposure, this particular problem is solved, and to all intents and purposes the atmosphere is 'frozen', so that the result is the same as it would be if there were no atmosphere at all. Unfortunately, a single very short exposure is of little use. Speckle interferometry involves taking a large number of short-exposure photographs ('snapshots', in fact) in quick succession, enhancing the images electronically, and then sorting them out by means of a computer. The end product is a high-resolution picture of the object. The larger the telescope used, the greater the resolution.

The results are surprisingly good, and it has become possible to attain resolving limits of below one-tenth of a second of arc. The apparent diameter of the Red Giant star Betelgeux, in Orion, is about 0.05 second of arc, corresponding to a real diameter of about 250 million miles. In 1975 astronomers at Kitt Peak, in Arizona, used the 158 in reflector to study Betelgeux by the speckle technique. They recorded bright and dark patches, and though there is now some doubt about the reality of these features the technique is at least promising.

Interesting results have been obtained with some types of galaxies. Messier 77 in Cetus, the Whale, is what is termed a Seyfert system, with a bright core and weak spiral arms. It is extremely active, and there have been suggestions that in its core there is a massive Black Hole. Speckle interferometry has confirmed that the core is indeed very small, so that it cannot be made up of stars scattered over a wide area. This does not prove the Black Hole hypothesis, but it does make it appear more plausible.

The technique is not restricted to objects at great distances; it is also of use in studies of certain members of the Solar System. In particular there are the asteroids or minor planets, most of which move in orbits between those of Mars and Jupiter. The largest member of the swarm, Ceres, has a diameter of less than 700 miles; the rest are much smaller, and appear as dots of light so that their diameters are very difficult to measure. Using speckle interferometry, astronomers at the European Southern Observatory in Chile have measured the apparent sizes of several asteroids. Juno, the third asteroid to be discovered (by the German astronomer Karl Harding, in 1804) turns out to have a longest diameter of 179 miles; this is probably accurate to within ten miles either way. The apparent diameter is only one-third of a second of arc. Juno is not regular in shape; its shorter diameter is no more than 143 miles.

Even more interesting is the case of Pluto, which is usually called 'the outermost planet' even though at the present time it is closer-in than Neptune, and its planetary status is very much a matter of debate. Photographic images indicated that it might be accompanied by a satellite, and the satellite was even given a name, Charon. Yet ordinary photographs could not show Pluto and Charon separately and many astronomers were decidedly sceptical about Charon's real existence.

The problem was cleared up by two French astronomers, D. Bonneau and R. Foy, using the powerful Canada–France–Hawaii reflector on the summit of Mauna Kea in Hawaii. Speckle interferometry showed that Pluto and Charon really are separate bodies. In 1982 further studies with the same technique were made at the European Southern Observatory, and removed any lingering doubts. The apparent separation between the two bodies is less than one second of arc.

Clearly, then, the new methods are of immense importance. Speckle interferometry is still in its infancy, but it has already shown that it has a great future — even when the 94 in Space Telescope is launched, which will probably be within the next few years.

33 THE HELIOSPHERE

On March 2 1972 NASA scientists launched their space-probe Pioneer 10 toward the planet Jupiter. In December of the following year it made its pass of the giant planet, sending back magnificent pictures as well as a vast amount of miscellaneous information. Originally it had been designed to operate for no more than 30 months — and if it had done no more, it would still have been regarded as an outstanding success. In fact, Pioneer 10 is still active today, and is proving to be more important than its makers had dared to hope.

By now it has reached the outer part of the Solar System. On April 25 1983 it crossed the orbit of Pluto, which at the moment is closer in than Neptune. It was not until Neptune's orbit too was crossed, on June 13, that Pioneer had definitely left all the planets behind. It will never come back; it has begun an endless journey between the stars, but we may hope to keep in touch with it at least until 1991, when it will be a full 5,000 million miles from the Sun. Its prime duty henceforth is to give us information about the vast, extended region known as the heliosphere — that is to say, the region of space in which the Sun's influence is dominant.

It has long been known that the Sun sends out streams of particles in all directions, making up what is called the solar wind. Its effects are detectable in many ways; for instance, the solar wind acts upon the tails of comets, 'pushing' them so that they always point more or less away from the Sun. (When a comet is receding from the Sun, it actually travels tail-first.) But the solar wind is not 'visible' in the ordinary sense; it comes from the region of the Sun's atmosphere known as the corona, which can be seen with the naked eye only on the rare occasions when the Moon covers the Sun completely, at a total solar eclipse.

The corona is at a high temperature of well over a million degrees. This is not the same as saying that it is 'hot'. Temperature is a measure of the speeds at which the atomic particles are moving around; the quicker the motions, the higher the temperature. In the corona, the speeds are very great, but there are so few particles that they do not produce much heat. (There is an analogy here with a firework sparkler and a red-hot poker. Each spark of the firework is at a high temperature, but has so little mass that the firework may safely be held in the hand — while I would certainly not recommend gripping hold of a glowing poker!)

The temperature of the corona ranges between one and five million degrees centigrade. This means that it emits X-rays. From the American Skylab station, the astronauts took X-ray pictures showing that in the corona there are denser 'active regions' which show up as bright patches, and are composed of material held in by closed loops of the Sun's magnetic field. There are also more rarefied regions, known as coronal holes. Solar wind particles flow out of the coronal holes, along the lines of magnetic force. Moreover, the Sun is subject to violent localised outbursts, which send out bursts of particles; these overtake the slower-moving particles of the regular solar wind, and produce shock-waves, so that the solar wind itself becomes gusty and variable. The average velocity of the solar wind past the Earth is almost one million miles per hour.

The Earth, of course, has a magnetic field, and is surrounded by zones of electrically-charged particles known as the Van Allen zones. When the solar wind is particularly strong, the particles from it overload the Van Allen zones and there is a cascade effect in the Earth's upper air, producing the lovely glows which we call auroræ or polar lights. In addition, the solar wind distorts the Earth's magnetosphere, squeezing it inward on the side turned toward the Sun and stretching it out into a long tail on the side turned away from the Sun. Therefore, the Earth's magnetosphere — the region in which our magnetic field is dominant — is pear-shaped.

Pioneer 10 investigated the magnetic field of Jupiter, which is much stronger than that of the Earth, and found the same effect; indeed the Jovian field can stretch out as far as the orbit of the next giant planet, Saturn. But what about the Sun's heliosphere, which may be said to end at the 'heliopause' where the solar wind can no longer be detected? Before the flight of Pioneer 10, it was believed that the solar wind would fade away not far beyond the orbit of Jupiter. We now know that it extends much, much further than that.

Here too we have a pear-shaped region, because the Sun is moving round the centre of the Galaxy at a high speed — around 44,000 mph. Space is not empty; it contains tenuous material, and there are magnetic effects too. Consequently, the heliosphere is compressed on the side in which the Sun is heading, and drawn out into a long tail in the opposite direction. It is kept inflated by the solar wind, and varies considerably in extent, because the Sun itself is not constant; it shows maximum activity every 11 years or so when, the solar wind is strongest and most gusty, so that the heliosphere is then at its largest.

There is another interesting effect here. This concerns cosmic rays, which are not really rays at all, but atomic nuclei, most of which come from deep space. When they reach the heliopause they have to penetrate it, and this is most difficult at the times of solar maximum, so that when the Sun is at its most energetic we receive fewer cosmic rays.

Pioneer 10 also showed that, contrary to expectation, the speed of the solar wind does not slow down with increased distance from the Sun. It had been thought that there would be a slowing-down process caused by collisions between the solar wind particles and the interplanetary medium, but nothing of the sort seems to happen. Another novel effect is that the shock-waves driven outward from the solar storms reverberate around the heliosphere for many months, and may even make it quiver in the manner of a gigantic jelly!

How far does the heliosphere extend? Pioneer 10 is heading 'downwind', and we may lose contact with it before the boundary is reached. However, its sister probe, Pioneer 11, launched a year later, is leaving the Solar System in the opposite direction, 'upwind', and here the heliopause is much closer, so that we may hope to maintain contact until the boundary is reached. Certainly astronomers hope so, because in this case we will be able to obtain information from true 'outer space', the regions which are not polluted by the solar wind.

The Pioneers were new-type probes in another way; they were not powered by solar panels. At the distances they were expected to reach, there would be too little sunlight; at the distance of Neptune, the strength of solar radiation is only about 1/1,000 of that which we receive on Earth. Instead, each Pioneer is powered by four units which make use of the decay of a radioactive substance (plutonium) to generate electrical power. The system has worked remarkably well.

Eventually, of course, we will lose all track of the Pioneers, and it is interesting to work out where they will go. Pioneer 10 is now travelling away from us at about 30,000 mph but in time the speed relative to Earth will drop to a steady 25,000 mph or so. It should reach the distance of the nearest star, Proxima Centauri (rather over four light-years) in 26,000 years. However, it is not moving in the direction of Proxima, and will not approach it to within six light-years. It will be only 3.3 light-years from the dim Red Dwarf Ross 248 in about 32,600 years; within seven light-years of Lambda Serpentis in 173,000 years, and then, in 227,000 years, it will be only 6.4 light-years away from a really bright star, Altair in Aquila, the Eagle.

Will Pioneers 10 or 11 ever be found by an alien civilization? The chances of any such thing are slight, but not nil — and just in case it happens, the Pioneers are equipped with plaques which will, it is hoped, give any finders a clue as to the planet of origin. Recently it has been suggested that such a procedure is dangerous in as much as it might lead unfriendly aliens in our direction. To me, this seems absurd. Any civilization capable of interstellar flight will have progressed so far in its evolution that it will have put the futility of war and conquest far behind it.

As to the eventual fate of the probes — well, it is difficult to say; presumably they will continue moving between the stars, unseen and unheard, until they are destroyed by collision with some solid object. They are our first, though not our last, messengers into interstellar space.

34 ILLUMINATING QUASARS

Radio astronomy began in the early 1930s, when a Czech-American radio engineer, Karl Jansky, made the fortuitous discovery of radio waves coming from the Milky Way. He never really followed up this pioneer work, and it was not until after the end of the war that radio astronomy became an all-important branch of science.

Early radio telescopes suffered from poor resolution; that is to say, they could not fix the positions of sources at all accurately. Yet it soon became clear that brilliant stars such as Sirius and Vega were not strong radio emitters. The most energetic sources were associated with much less striking objects. Old supernovæ accounted for some of them, and some galaxies were strong at radio wavelengths, but other sources were much more puzzling, and did not appear to be associated with anything at all. Many of these were contained in the third Cambridge catalogue of sources, compiled in the early 1960s. One of them, 3C-273, lay in the constellation of Virgo, and was one of the most powerful of all.

Then came a stroke of luck. 3C-273 is so placed that it can be hidden or occulted by the Moon. This happened on August 5 1962, and was observed by the Australian radio astronomers Hazard, Mackey and Shimmins. Of course the radio waves were cut off as soon as the Moon passed over the source; since the position of the Moon at this moment was known, the exact position of 3C-273 could also be found. The information was sent to the Palomar Observatory in California, and a visual search was made. The source seemed to coincide in position with a faint blue star of magnitude 13 — but when Maarten Schmidt used the 200 in Hale reflector to examine the spectrum of the object, he was taken by surprise. 3C–273 was not a star at all, but something much more dramatic. It was the first quasar, a convenient abbreviation of 'quasi-stellar radio source'.

Stellar spectra may show bright emission lines and dark absorption lines. If the lines are shifted toward the long-wave or red end of the spectrum, as compared with spectra obtained in the laboratory, the inference is that the object is receding — and the greater the red shift, the greater the speed of

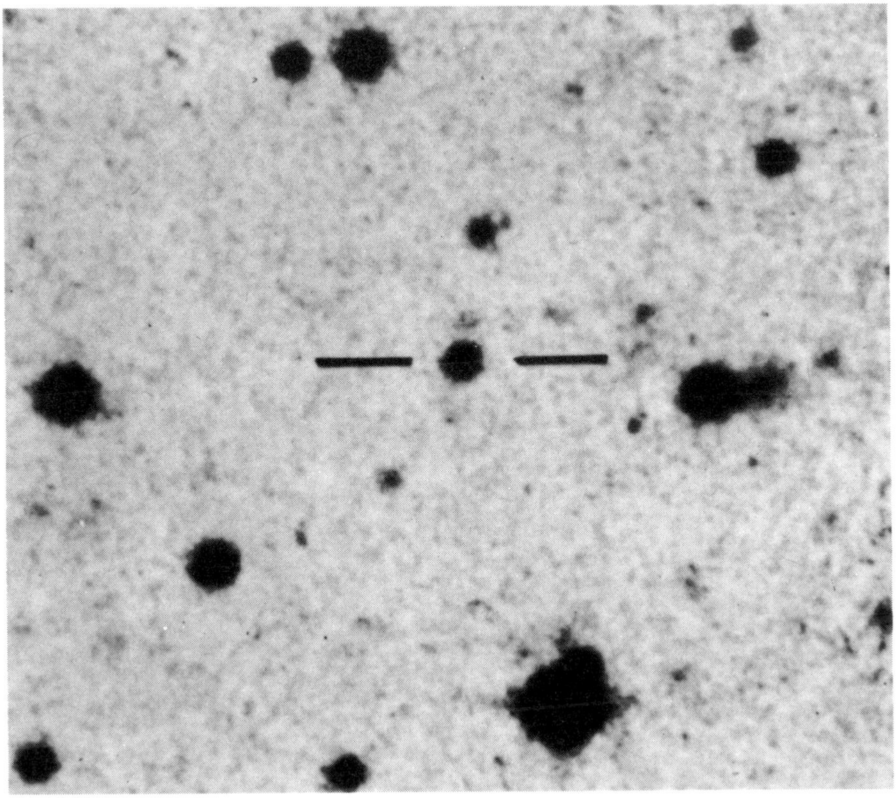

The Quasar PKS 2000–330. The quasar lies between the identification bars. Photograph from Siding Spring in Australia. This is the most remote object so far discovered.

recession (this is, of course, the well-known Doppler effect). 3C-273 showed emission lines so strongly red-shifted that they indicated a velocity of well over 20,000 miles per second, which meant that the distance must be of the order of 2,500 million light-years. It followed that the quasar must be incredibly luminous, even though it looked very small compared with a galaxy.

More and more quasars were found, almost all of them with still greater red shifts. Some of them seemed to be more than a hundred times as luminous as the most powerful galaxies, and they were variable over periods of a few days or even a few hours, so that they could not be very large by cosmical standards. Astronomers had to explain how so much energy could be contained in a volume of space no larger than the Solar System. Various theories were proposed (chains of supernovæ, the mutual annihilation of matter and anti-matter, and so on), but none seemed to be really convincing.

Another curious phenomenon became evident. As well as the bright lines, the spectra of quasars showed dark lines, but these had different, smaller red shifts, and altogether the spectra were so confusing that many investigators began to doubt whether the red shifts were Doppler effects at all, in which case there was no need to assume the quasars to be so remote and so luminous. Now, however, an explanation has been found by Alec Boksenberg, the Director of the Royal Greenwich Observatory, and his colleagues. The dark lines do not belong to the quasars, but to clouds of material lying between the quasars and ourselves. Each cloud produces its own set of absorption lines, and these are superimposed on the spectra of the quasars. This was a particularly useful discovery, since it meant that we could learn something about the intergalactic clouds themselves.

Not all quasars are radio sources. Many show 'jets' of material, and we receive strong radio emissions only if we are so placed as to be looking almost along the axis of the jet.

By now, it seems almost certain that a quasar is nothing more nor less than the nucleus of an active galaxy, probably with a Black Hole in its centre. A Black Hole is the region round an old, collapsed star where the gravitational pull is so great that not even light can escape from it; therefore, the region is virtually cut off from the rest of the universe. (I appreciate that this is a gross over-simplification, but it will serve for the moment.) A Black Hole has been called a cosmic cannibal; it pulls in material from around it, and before the material disappears forever into the Hole it is intensely heated, so that it radiates fiercely.

If the central Black Hole of a quasar is very small, it will not make its presence felt. If it is very large, it can swallow up whole stars 'at a gulp', and again the effects will not be noticeable. A medium-sized Black Hole will act differently. As the unfortunate star spirals in toward the boundary, it will be literally torn to shreds, and a large fraction of its mass will be converted into radiation before the star is completely destroyed and engulfed. This would explain the energy-sources of the quasars which are particularly powerful.

It is fair to say that although this explanation is now very popular, there are some astronomers (notably Sir Fred Hoyle) who do not agree with it. In America, Halton Arp has drawn attention to groups of objects which are sufficiently closely aligned to be regarded as genuinely associated with each other; both galaxies and quasars are included, and their red shifts are different. More researches are needed, but at the moment the majority view is that the quasars really are immensely remote and super-luminous.

No quasars lie anywhere near the Solar System. The most remote quasar so far discovered is PKS 2000-330, found from the Parkes Radio Astronomy Observatory in New South Wales. It is thought to be about 13,000 million light-years away, and to be receding at 95 per cent of the velocity of light. Yet PKS 2000–330 is by no means the faintest of the quasars; if it lay at a still greater distance it could be detected without much difficulty.

Evidence is growing that the quasars are not uniformly distributed. 3C-273 is one of the closest. Remember, when we look at an object which is (say) 10,000 million light-years away we are seeing it as it used to be 10,000 million years ago, and if the quasars lie within certain reasonably-defined limits of distance it follows that they were active at around the same time.

Does every galaxy go through a quasar stage at some period in its evolution? And is there a Black Hole at the core of our own Galaxy? So far, we have to admit that we do not know. Neither have we any precise knowledge of the extent of the observable universe. The further away a galaxy or a quasar lies, the faster it is receding, so that the entire universe is expanding. If this rule holds good, we will eventually come to a distance at which an object is racing away at the full velocity of light, in which case we will be unable to see it. Estimates of this critical distance range between 15,000 million and 20,000 million light-years, with a slight preference for the lower value.

Because quasars are so powerful they can be seen over distances from which ordinary galaxies would be undetectable, and this is one reason why they are so important. They can be used to study closer objects, and may be said to 'illuminate the background'. By now we have found out a good deal about them, but there is still much that we do not know.

35 NEWS FROM VEGA

One of the brightest stars in the sky is Vega, or to give it its official designation, Alpha Lyræ. It is almost overhead during summer evenings in England, and cannot be mistaken, partly because of its brilliance and partly because of its beautiful blue colour. It is hotter and more luminous than the Sun; is 26 light-years away, which works out to about 150 million million miles. Its luminosity is between 50 and 60 times that of the Sun. By cosmical standards it is fairly youthful. Our Sun is at least 5,000 million years old, but Vega cannot date back more than 1,000 million years, and 350 million years is a more likely estimate.

Powerful though it is, Vega is by no means exceptional; Deneb, also high in the evening sky during summer, can equal at least 60,000 Suns, but it is much further away, and does not appear so prominent. In fact Vega has always been regarded as an ordinary star, conspicuous in our skies because on the cosmical scale it is fairly close to us. Now, suddenly, all this has been changed, due to some startling information sent back by IRAS, the Infra-Red Astronomical Satellite.

IRAS, launched in January 1983, showed at an early stage that it was going to be a great success. The data from it is received at the Rutherford-Appleton Laboratory in Oxfordshire, and it was here that two of the American scientists on the team, Dr Hartmut ('George') Aumann of the Jet Propulsion Laboratory at Pasadena in California, and Dr Fred Gillett of the Kitt Peak Observatory in Arizona, made their remarkable discovery. They were using various stars as sources for calibrating the infra-red telescope in IRAS when Dr Gillett said, suddenly: 'Hey, Alpha Lyræ has a huge excess!' Much more infra-red radiation

Left *Dr Aumann and Dr Gillet at Rutherford-Appleton Laboratory, 1983.*

Below right *Radio telescope at Rutherford-Appleton Laboratory, Oxfordshire, used for receiving the IRAS results.*

was being received than had been expected, and further tests showed that the radiation really was associated with Vega; all the other stars being used for calibration were completely normal. The Vega source was about 20 seconds of arc in diameter, and as the equipment was accurate to within two seconds there seemed to be no possibility of mistake. The next step was to try to find an explanation, Dr Aumann commented: 'Maybe, we thought, it is due to mass flowing out from Vega'. But that possibility was soon ruled out of court. The material was not flowing away from Vega, and clearly it had been there for a long time.

Further studies showed that the material took the form of a cloud of particles, many of them much larger than the 'dust' particles found in interstellar space. The infra-red radiation came from a region extending out to about 7,400,000 miles from the star, which is about 80 times the distance between the Earth and the Sun, so that there had to be a considerable quantity of it. 'I started to think about it,' Dr Aumann told me. 'If there are small particles round Vega, there must be large particles also. Very careful analyses have been made of what are called fragmentation products. When I applied these methods in the case of Vega, I was astonished because it seemed that the total mass was much the same as that of all the planets in our Solar System combined.'

As the analyses went on, the situation became even more fascinating. Our

Solar System is believed to have been born from a cloud of material associated with the early Sun — a solar nebula, in fact — and it seemed that the same process might now be going on around Vega. Dr Aumann and Dr Gillett came to the conclusion that the material must contain bodies much larger than dust-grains, because very small particles would already have been drawn back into the star. This would leave intermediate and large-scale débris in orbit, and this débris could already have developed into planets. The temperature of the material was estimated as -184°C, which is about the same as the temperature of the icy particles making up the inner rings of Saturn.

The discovery is of the greatest significance, but we must be careful not to take it too far. Vega, remember, is younger and more powerful than the Sun. It has had much less time to evolve, and any planetary system will therefore be in an earlier stage than our own. Life is notoriously slow to develop, and it is hardly likely that any intelligence exists in the system of Vega, even assuming that there really are bodies of planetary size. Then, too, Vega sends out much more short-wave radiation than the Sun, and this could be harmful to living things unless they were well protected by a suitable atmosphere. As yet we depend entirely upon the infra-red findings, and direct observation of the 'cloud' will be far from easy. Faint, cool material is effectively drowned by the brilliance of Vega itself.

It is interesting to speculate about what the system of Vega would be like if we could see it from close range. Dr Gillett pictures 'a number of rings, in the stages of coalescing and possibly forming planets', but he was quick to add that

this was really nothing more than controlled imagination, because it is not possible to derive such detail from the IRAS results.

Perhaps the most important point of all is that if Vega, a normal star, can produce planets it is more than likely that other stars will evolve in the same way. Planetary systems will then be very common in the Galaxy, and the Sun's family cannot be regarded as anything in the way of a cosmic freak.

What will happen to the Vega system in the future? So far as the development of life is concerned, the prospects do not appear to be good, because Vega is running through its life-cycle at a much faster rate than our sedate Sun. The Sun is not likely to change drastically for at least 5,000 million years in the future, but by that time Vega will no longer exist in its present-day form; it will have passed through its Red Giant stage, with disastrous results to any planets moving round it. Dr Aumann was quite positive. 'We very much doubt that Vega's system will ever evolve into a Solar System like our own,' simply because it will not have enough time. Vega can hardly exist in its present-day form for more than a few hundreds of millions of years.

This, so far, is about as far as information and speculation can take us. Meanwhile, look up at Vega on the next dark, clear night when it is high above the horizon. You will see it shining down with its lovely bluish radiance — and it is enthralling indeed to reflect that it, like the Sun, may well be the centre of a family of planets.

36 COMETS WITH DUSTY TAILS

At the moment there is a great deal of interest in comets. This is due largely to the return of Halley's Comet, which is due to reach its closest point to the Sun in February 1986. Meanwhile, some new information about fainter comets has come from IRAS. This remarkably successful satellite has discovered five new comets. The first of them was detected independently by two ground-based observers, Araki in Japan and George Alcock in England, so that it is referred to as Comet IRAS-Araki-Alcock. The others belong to IRAS alone.

These discoveries may be much more important than might be thought, because apparently faint comets are much more plentiful than has been generally believed. IRAS has also been used to detect another new feature — a long, dusty tail to a known comet, Tempel 2.

The comet was originally discovered by the German astronomer Ernst Tempel in 1873. It has a period of $5\frac{1}{3}$ years, and is well known; it has been seen regularly, and the present return is the 17th to be observed. It is always too faint to be seen in telescopes of the size generally used by amateurs, and even in large instruments it appears as nothing more than a dim blur of light. Nothing could be more unlike the brilliant comets so often seen in paintings and photographs; but it is fairly typical of the family of short-period comets.

Comets are of two definite types. The periodical comets move round the Sun

West's comet, photographed on March 9 1976 by Paul Doherty. The comet had a long tail, and was a bright object. It has now disappeared, and will not return for thousands of years.

in orbits which bring them back to perihelion (that is to say, closest to the Sun) every few years. Encke's Comet, for instance, has a period of only $3\frac{1}{3}$ years. But the only bright comet with a period of less than several centuries is Halley's (76 years) and the spectacular visitors which are seen now and then have periods so long that to all intents and purposes we may regard them as non-periodical. It is usually believed that at an immense distance from the Sun there is a whole swarm or reservoir of comets, and that occasionally one of these is perturbed and sent hurtling sunward. If it then comes under the gravitational influence of one of the giant planets (usually Jupiter) it may be twisted into a short-period orbit.

Each time a comet nears the Sun and is strongly heated the icy materials in its head begin to evaporate, thereby forming a tail. The comet loses some mass at each perihelion, which is why comets are relatively short-lived members of the Solar System; several periodical orbits which were followed regularly during the 19th century have now disappeared. Tempel 2 is a typical short-period comet, and up to now there has been no suggestion of a tail.

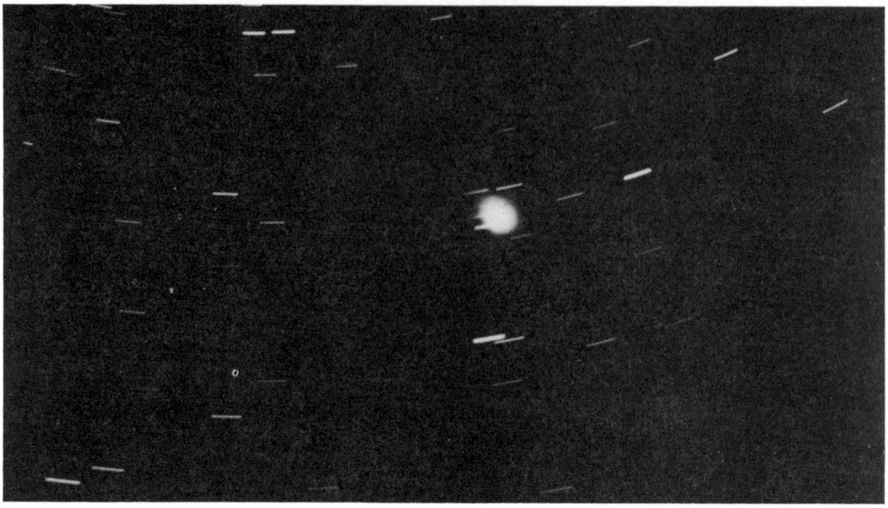

Comet IRAS–Araki–Alcock.

Some months ago, IRAS results were being examined by an investigating team at Leicester University. They picked up what appeared to be a whole group of small particles, moving through space in parallel paths. There were immediate signs that these particles might be associated with a comet, and before long infra-red radiations were detected from Tempel 2 which was in just the right place at just the right time.

As the investigations went on, some very unexpected facts emerged. Tempel 2 does indeed have an extremely long tail streaming away from its head. It is made up of what may be called 'dust', released from the central part of the comet under the influence of the Sun. The tail sends out very little visible light (too little, indeed, for us to detect it at all) but it is quite obvious in the infra-red range. Fortunately, the comet was placed virtually 'edge-on' at the time of the discovery, so that the IRAS telescope was looking through the thickest part of the dust-tail. If this had not been so, it is unlikely that the tail would have been found.

Clearly it was important to find out whether all faint periodical comets had similar tails. IRAS was therefore commanded to turn its attention to another regular visitor — Tempel 1, also discovered by Ernst Tempel. It was seen in 1867, 1873 and 1879, after which it was lost until it was recovered in 1967. The period is $5\frac{1}{2}$ years. IRAS detected no tail, which may mean either that the instruments were not sensitive enough, or that Tempel 2 is exceptional in having a dust-tail of this sort.

No doubt time will tell. At least it is significant to learn that as well as paying attention to objects many light-years away, IRAS was also of great value in improving our knowledge of these faint, ghostly members of the Solar System.

37 THE SKY FROM MOUNT WILSON

Our programme from Mount Wilson was filmed in July 1983, and broadcast in the following October. Before I came to revise the article for this book there had been some unfortunate developments. For obvious reasons, therefore, I have made important modifications to the original text.

Overlooking the city of Los Angeles is the moderately high, rather beautiful peak of Mount Wilson. At its summit there is a dome, or rather several domes, one of which contains the telescope which was for many years the largest in the world. From its completion in 1917 until it was finally surpassed in 1948, the 100 in Hooker reflector at Mount Wilson was in a class of its own, and it led to a complete revolution in all our ideas about the universe.

It was master-minded by an energetic and resourceful American astronomer, George Ellery Hale, who first achieved fame in 1892 when he invented an

The Snow solar telescope at Mount Wilson. I agree that it looks more like an old shed than a telescope, but it is still extremely effective!

instrument known as the spectroheliograph, used for studies of the Sun. At first Hale was mainly a solar observer, but he was also very alive to the need for building large telescopes to collect as much light as possible from dim, remote stars and star-systems. In fact his constant call was for 'More light!' and his first real achievement in this direction was to persuade a millionaire, Charles T. Yerkes, to finance a refracting telescope with a lens 40 in in diameter. The 40 in at the Yerkes Observatory, 70 miles from Chicago, remains the largest of its kind, and will probably never be outranked. Lenses have to be supported round their edges, and if they become too large and heavy they distort under their own weight. (A 49 in was later made in France, but proved to be a total failure.) A mirror, on the other hand, can be supported on its back, because the light does not have to pass through it; and Hale realized that the real future lay with reflecting telescopes.

He approached the Carnegie Foundation, named after Andrew Carnegie, one of the few men as rich as Yerkes. Funds were made available for a 60 in reflector and George Ritchey, the leading mirror-maker of the time, was put in charge of the optical work. Obviously, the choice of a site was all important. To avoid the worst of the disturbances in the Earth's atmosphere it is advantageous to 'go high', and there were two peaks in California which seemed to be suitable: Palomar and Mount Wilson. Actually Palomar was the better, because it was further away from any city lights, but it was also much more difficult to get at and so the choice fell upon Mount Wilson. The 60 in reflector came into use in 1908, and proved to be an immediate success.

Even earlier, Hale had been responsible for setting up solar equipment on Mount Wilson. The first of these was the Snow telescope, so named because the money for it was provided by a Miss Snow. It was of unusual design; the sunlight was caught by a rotatable mirror or cœlostat, and sent horizontally on to a second mirror before being diverted into the analytical equipment. Since the Sun's rays arrived in a fixed direction, the heavy equipment did not have to be moved at all, which was a major advantage. The Snow telescope is still in use, but it does suffer from the defect of being affected by the warmth of the ground, and the next solar telescopes were of the 'tower' type, with the cœlostats high above the ground — 150 ft in the case of the largest instrument. It was from Mount Wilson, in 1908, that Hale proved the existence of strong magnetic fields associated with sunspots. The importance of this discovery in solar physics can hardly be over-estimated.

Though the Sun is so close to us on the cosmical scale, it would be wrong to claim that our knowledge of it is at all complete, and in fact we may know less than we fondly imagined a few years ago. There is still considerable argument about the basic cause of sunspots and about the 11 year solar cycle. Every 11 years or so the Sun is very active, with many spot-groups and associated phenomena; activity then dies down, and at solar minimum the disk may be spot-free for many consecutive days. The last maximum was that of 1980, so that the next may be expected around 1991, though the cycle is not perfectly regular. There seems little doubt, too, that between 1645 and 1715 there were

The 60 ft solar tower at Mount Wilson.

very few spots, so that the cycle was suspended — for reasons which are, as yet, quite unknown. Working with the Mount Wilson solar telescopes, Dr Robert Howard has been studying large-scale phenomena. For example, there is a strong possibility that the interior of the Sun spins round more quickly than the surface rotation period of about a month.

Even before the 60 in had been completed Hale was planning a still larger telescope, this time with a 100 in mirror. Again he approached a friendly millionaire, this time John D. Hooker; Hooker promised the sum of 45,000 dollars which would pay for the mirror though not for the mounting. Casting the glass disk was a major problem, and to shape the mirror to the correct optical curve took Ritchey and his team six years. During the work, Andrew Carnegie visited Mount Wilson and was so impressed that he agreed to pay for the

Above *Dome of the 60 in, the first large telescope on Mount Wilson.*
Right *The 100 in Hooker Reflector at Mount Wilson.*

mounting. By November 1917 all was ready, and Hale and three companions turned the great telescope toward the planet Jupiter.

It must have been a traumatic experience. All they could see was a shimmering blur, with no definition at all. Could the mirror be faulty? Hale did not believe so, and when he went back to the dome, hours later, he looked at the blue star Vega and saw it as a beautifully-defined sharp point. The mirror had needed time to cool down after the daytime heat.

Before long the tremendous light-grasp became evident. The Hooker reflector could see stars far fainter than any previously observable, and in 1923 Edwin Hubble took advantage of this to make a classic discovery. Up to that time it was not known whether the strange objects known as spiral nebulæ were contained in our star-system or Galaxy, or whether they were independent galaxies, so distant that their light would take millions of years to reach us. With the 100 in, Hubble detected certain special variable stars in some of the spirals. These stars, known as Cepheids, brighten and fade regularly over

periods of a few days or weeks, and their behaviour gives a clue to their real luminosities. Hubble was able to measure the distances of the Cepheids merely by observing them, and it was clear that they were much too distant to be members of our Galaxy. The brightest of the spiral nebulæ is Messier 31, in Andromeda, which is just visible with the naked eye as a dim blur on a clear night. From the Cepheids in it, Hubble estimated its distance as 900,000 light-years, later amended to 750,000 light-years. Actually this is much too little, because of an error in the Cepheid scale which was not discovered until much later; the true distance of Messier 31 is 2,200,000 light-years. But Hubble had made the vital breakthrough, and no telescope but the Hooker reflector would have been adequate.

Until 1948 the 100 in reigned supreme. Then, again at Hale's instigation, the 200 in reflector was set up on Mount Palomar. It is sad that Hale did not live to see it (he died in 1938) but it is surely fitting that the 200 in should be known as the Hale reflector.

Since then the 100 in has been used for many other important researches. For example, Dr Olin Wilson, who joined the Observatory staff as long ago as 1931, has been carrying out investigations into the activity cycles of other stars. Dr Douglas Duncan has been using the 60 in to investigate the ways in which the stars rotate. It now seems that some stars are much more active than the Sun, perhaps by a factor of at least 30, and their rotation periods range between a week or so and several months. Generally speaking it is the younger stars which show the greatest activity, which is hardly surprising.

For this kind of work Mount Wilson is still ideally suited. The telescopes are in perfect order, and can now, when need be, adapt to modern type electronic devices which are superseding the sensitive plate, just as the camera superseded the human eye almost a century ago.

Yet a cloud of a different kind has been hanging over the Observatory. Conditions on Mount Wilson are not nearly so good as they used to be, because Los Angeles has spread, becoming not only larger but also much brighter; the skies around it are less dark than they were in Hale's day and this, of course, is a serious handicap. Moreover, the 100 in reflector completed nearly 70 years ago, is old-fashioned as compared with the great reflectors of today. On June 29 1929 the Carnegie Institute of Washington, which operates Mount Wilson, announced that during 1985 the Hooker reflector will be mothballed, and there will be financial reductions in support for the famous 150 ft and 60 ft solar tower telescopes (though not for the 60 in reflector, which has independent support). If carried through, this policy will mean the end of Mount Wilson as a leading astronomical institution.

It seems a tremendous pity, and we can only hope that something can be done. The Hooker 100 in is, after all, a very special telescope. It was in this dome, in 1923, that Edwin Hubble obtained the evidence showing that our Galaxy, with its hundred thousand million stars, is only one of many, and also that the universe is in a state of expansion. This alone is enough to ensure that Mount Wilson will be remembered as long as scientific history lasts.

38 THE SKY IN INFRA-RED

IRAS, the Infra-Red Astronomical Satellite, had an active lifetime of 11 months. It ceased transmitting in November 1983, but by then it had been responsible for a great many important pieces of research. Its first task was to undertake what may be termed a mini-survey of selected parts of the sky. During this operation 8,000 infra-red sources were found and later a much wider survey revealed 200,000 sources, many of which had been previously unknown. They are of many different types; some in the Solar System, others in the Milky Way Galaxy, and yet others in remote galaxies so far away that their radiations take millions of years to reach us.

One discovery was that of a strange body which may be either an asteroid, or else the remnant of a comet which has lost all its dust and gas. On October 11 1983 Simon Green, a member of the IRAS team, noted a rapidly-moving object which was seen in seven consecutive orbits of the satellite. It was quickly confirmed and photographed by the American observatories at Flagstaff and Palomar, and found to be very remarkable indeed. It is about 1.2 miles in diameter, and appears as a dim object of the 16th magnitude; it has an elliptical orbit which carries it to within nine million miles of the Sun. It takes 1.5 years to complete one orbit, and its path is inclined to that of the Earth by 20°. It has been named Phæthon.

Can it be a minor planet? The large asteroids keep to that part of the Solar System between the orbits of Mars and Jupiter, but there are smaller asteroids which cross the orbit of the Earth. These are known as Apollo asteroids, after the first-discovered member of the class. Up to now only one of them, Icarus, has been known to venture inside the orbit of Mercury; but Icarus can never approach the Sun much within 17 million miles, so that Phæthon is much more extreme. When near perihelion, it must be red-hot.

It has often been suggested that Apollo asteroids begin their careers as comets, so that they now represent dead comets from which all the dust and gas has been stripped away. It is quite definite that meteors are cometary debris, and many of the annual meteor showers are associated with comets — for instance, the autumn Orionid meteors are linked with Halley's Comet. The significant fact about Phæthon is that its orbit is very nearly the same as that of the Geminid meteors seen each December, so that it may be the defunct partner of the Geminid stream.

IRAS has also found three narrow, continuous rings of dust in the Solar System. The main ring lies in the asteroid belt (between 200 and 300 million miles from the Sun) and could be due to collisions between the asteroids themselves; its temperature is between -123°C and -173°C. There are also rings of dust about 9° above and 9° below the main ring. The existence of these features is not nearly so easy to explain. It has been suggested that they are due to collisions between comets and asteroids, but further researches are needed before we can be at all certain. The mass of the dust in the three rings is not great; it is equivalent to the mass of a rocky asteroid about 0.6 miles in diameter.

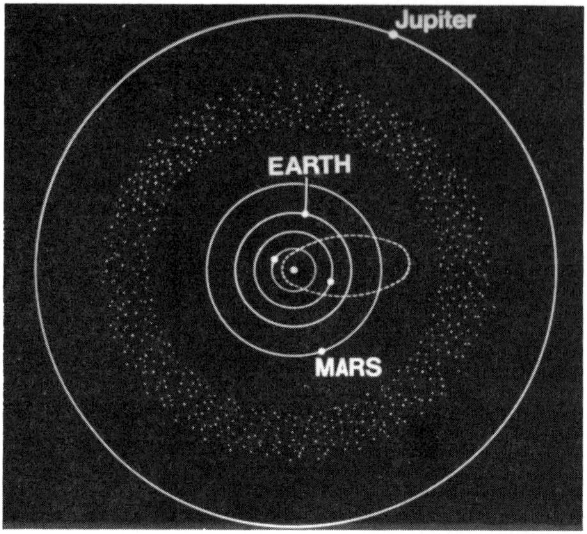

Orbit of the remarkable asteroid Phæthon, discovered by IRAS.

Other IRAS discoveries include the cool, possibly planet-forming material around Vega and other stars, and material associated with Betelgeux, the Red Supergiant in Orion. Betelgeux is an old star, and it is huge — its diameter is greater than that of the Earth's orbit round the Sun — so that it is not the sort of star which would be expected to have a system of planets. The material detected by IRAS is not a ring. It seems to have been 'left behind' Betelgeux as the star moves through space, so that it is found on one side of Betelgeux only.

One very important part of the IRAS programme concerns very young stars. These are surrounded by dusty cocoons, which of course are best detected in infra-red. It has been found that many of the relatively nearby dark clouds in the Galaxy contain embryo stars which are just about to begin their luminous careers. One such object has been found inside the dark cloud known as Barnard 5 in honour of its discoverer, the American astronomer Edward Emerson Barnard. At the moment it is sending out about ten times as much energy as the Sun, though its temperature is only about 220°C. Eventually it will turn into a star very like the Sun; the IRAS results show that solar-type stars go through stages of high luminosity during the first hundred thousand years of their existence. In another cloud, in the constellation of Chamæleon (too far south to be visible from England) 17 objects of similar type were found; they may be embedded inside the cloud, or they may be proto-stars which are clear of the cloud but still inside their dusty cocoons. Of course the famous Orion Nebula has come in for special attention; it has long been known to be a stellar nursery and fresh stars are condensing out of the nebular material.

Another discovery is that of wispy clouds of dust scattered widely over the sky. This has been termed 'infra-red cirrus', and may consist of graphite (carbon) particles formed in the outer atmospheres of stars and heated by starlight. IRAS has also looked at the centre of the Galaxy, around 30,000 light-years away in the direction of the lovely star-clouds in Sagittarius. We cannot see the central regions visually, but the infra-red radiation can come through to us. The centre of the Galaxy is a mysterious place, and IRAS will certainly have helped us to a better understanding of it.

Beyond our Galaxy there are many other galaxies, some of which are spiral in form while others are elliptical or irregular. Generally speaking the spirals are the most prominent in infra-red, because they still contain material which has not yet been formed into stars, whereas in the elliptical galaxies most of the star-forming material has been used up. Some galaxies which are very dim visually have been found to be very powerful at infra-red wavelengths.

Quite apart from all this there are also infra-red sources, some of them strong, which do not appear to be associated with anything at all. The positions of these sources can be pinpointed very accurately, but nothing can be detected either visually or at radio wavelengths. As yet they remain unexplained. They could be inside the Milky Way Galaxy, so that they could be stars which have not yet evolved far enough to produce any visible light; they could be old stars concealed by thick shells of surrounding material; they could even be external galaxies too faint to be recorded in any other way.

All in all, IRAS must be regarded as one of the most successful of all artificial satellites to date. It has done more than its makers had even hoped, and its 11 months of active life have been immensely fruitful.

39 WEIGHING A BLACK HOLE

Ever since their discovery, more than 20 years ago now, quasars have presented astronomers with all manner of problems. They are immensely distant and immensely powerful; a single quasar may shine as brilliantly as a hundred complete galaxies — and a galaxy such as our own contains 100,000 million stars. Moreover, quasars vary in light, sometimes in a matter of days. This means that they cannot be large. Otherwise, light coming from the far side of the object would lag behind light coming from the near side, and the result would be apparently steady emission.

Gradually, it emerged that the most likely power-source was a Black Hole in the centre of the quasar, while the quasar itself was the nucleus of an active galaxy. A Black Hole is produced by an old, massive star which has collapsed in its old age, and is now pulling so strongly that not even light can escape from it. There are also the so-called BL Lacertæ objects, which resemble quasars but are less luminous. Finally there are Seyfert galaxies, which were first identified by Carl Seyfert as long ago as 1942; they have very bright, condensed centres and only weak spiral arms. If far enough away, a Seyfert galaxy would look like a quasar.

Seyferts are less powerful than quasars, but it now seems that each Seyfert has

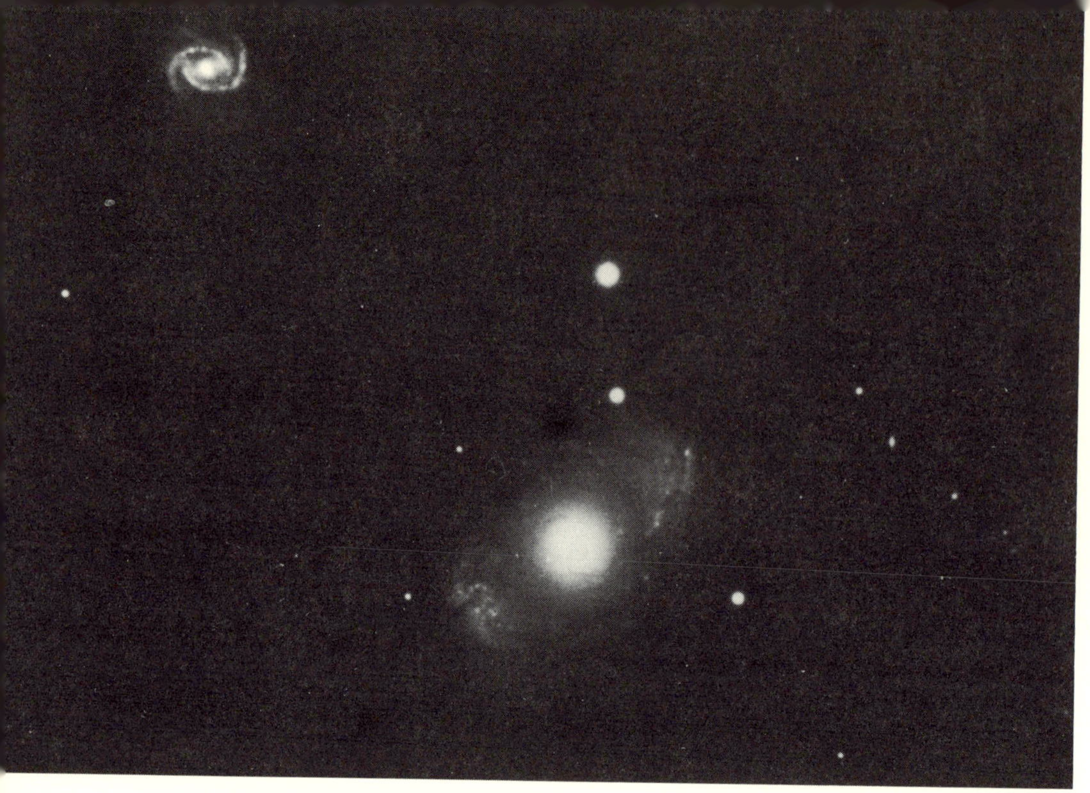

a 'mini-quasar' in its centre. And this gave astronomers at the Royal Greenwich Observatory the idea that we might be able to give the first reliable values for the mass of a Black Hole.

The galaxy which provided the best opportunity was a Seyfert, known as NGC 4151, in the northern constellation of Canes Venatici, the Hunting Dogs. It is not bright enough to be seen with small telescopes (in the 15 in reflector at my observatory in Selsey it shows up as a faint blur), and it is some 50 million light-years away.

Like many Seyferts, it varies in light, and the variation comes from the nucleus of the galaxy, where the Black Hole presumably lies. Investigations carried out by the artificial satellite IUE (International Ultra-violet Explorer), which was equipped with an 18 in telescope suited to observing in the short wavelengths, showed that when the nucleus brightens, the effects spread out into the spiral arms of the galaxy. But — and this is the important point — there are delays; obviously the outburst reaches the nearer spiral arms first. The ultra-violet spectra of the separate arms can be studied, and the times when the brightenings are observed shows how far the arms are from the centre of the outburst.

This is not all. We can also make use of the well-known Doppler effect to find out just how the material in the spiral arms is moving. If the light we receive is slightly 'too red', the source is receding; if it is 'too blue', the source is approaching. The actual colour-change is generally too slight to be noticed, but

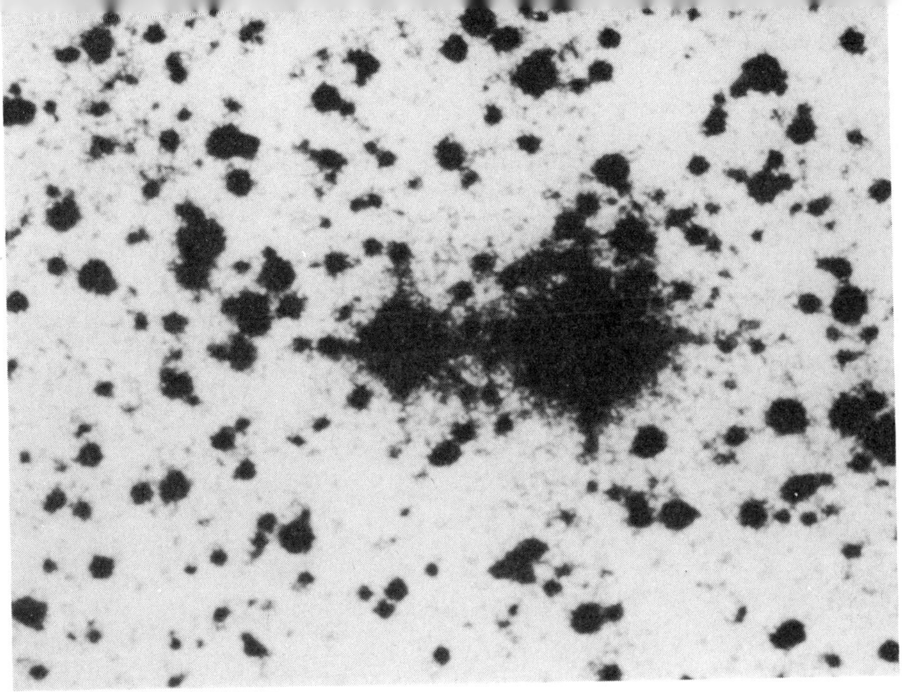

Above *Cygnus X–1, the system which may contain a Black Hole.*
Left *A Seyfert galaxy, Messier 77 in Cetus. It has an active, condensed nucleus. A smaller spiral, NGC 1055, is seen to the upper left.*

it shows up in the shifts of the spectral lines. Therefore we can tell how fast the material is moving, and how far it lies from the Black Hole at the centre of the galaxy. The situation is basically the same as with the planets of the Solar System. The nearer a planet is to the Sun, the faster it moves; thus the Earth is travelling faster than Mars, but slower than Venus. The ways in which the planets move give a reliable clue to the mass of the body controlling them. With the Solar System, the controlling mass is the Sun. With NGC 4151, the controlling mass is the central Black Hole.

Two clouds inside NGC 4151 were studied in detail. One, 13 light-days from the nucleus, was moving at 8,700 miles per second; the other, at 30 light-days, at 6,800 miles per second. From this, the Greenwich team deduced a value for the mass of the central Black Hole between 50 million and 100 million times that of the Sun.

It all sounded very convincing, but then, after the publication of the paper in the *Monthly Notices* of the Royal Astronomical Society, two amateur astronomers independently found that something was badly wrong. R. F. Wood, of the British Astronomical Association, and Dr Y. N. Chen of Winterthur in Switzerland made some calculations, and derived a mass for the Black Hole not of 50 million to 100 million times that of the Sun, but 500 million to 1,000 million times.

These results were communicated to the British Astronomical Association, which passed them on to the Greenwich team. It did not take long to confirm that

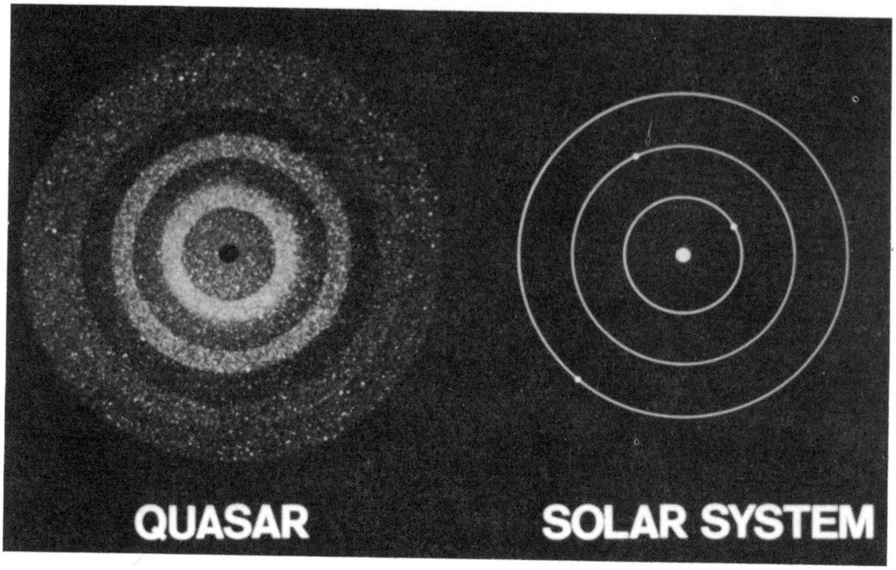

How to weigh a Black Hole — by comparing the situation with that in the Solar System.

the amateurs were absolutely right. It had been a case of a simple mathematical slip; the value of the gravitational constant had been incorrectly inserted, so that the final results were too small by a factor of ten. Somehow or other, the mistake had slipped past not only the authors of the original paper, but also past the referees to whom it had been submitted. This, surely, is yet another proof that the work of the amateur astronomer is not to be despised!

Staggering though it may seem, the revised value for the mass of the Black Hole is probably at least fairly reliable. The quasars, and the mini-quasars inside Seyfert galaxies, have started to yield up some of their jealously-guarded secrets.

40 FOMALHAUT: ANOTHER PLANETARY SYSTEM?

Fomalhaut, the leader of the small southern constellation of Piscis Australis or Piscis Austrinus, the Southern Fish, is not one of the brightest stars in the sky. It comes 18th in order, and with an apparent magnitude of 1.2 it is well over a magnitude fainter than Arcturus or Vega. Moreover, it is never seen to advantage from the British Isles. During autumn evenings it appears low down over the southern horizon, directly below the Square of Pegasus; the two 'right-hand' stars of the Square, Skat and Markab, point directly toward it which is probably the best way to identify it. From North Scotland it barely rises, and is extremely difficult to see at all.

From countries such as South Africa and Australia, Fomalhaut passes over the zenith, directly above the observer, and is easy to locate because there are no other bright stars anywhere near it. The Southern Fish itself is a very dim, obscure constellation, and despite the presence of Fomalhaut there seems no reason for it to have a separate identity, particularly as no mythological legends are attached to it. The name 'Fomalhaut' is Arabic; an alternative Arabic name seems to have been Al Difidi al Awwal, or the First Frog! In Syrian and Caananite lands it was honoured as a symbol of the sea-god.

Fomalhaut has a faint companion 2° to its south. However, the companion is only of the 14th magnitude, much too faint to be seen in small telescopes. There is a chance that it is genuinely associated with Fomalhaut, but the distance between the two is almost a light-year, so that there may be no real connection.

Fomalhaut itself is one of our nearest stellar neighbours; its distance is 22 light-years, corresponding to approximately 130 million million miles. This may seem a long way, but of the brightest stars in the sky only Alpha Centauri, Sirius, Procyon and Altair are closer. Fomalhaut is pure white, with a luminosity 13 times that of the Sun; its diameter is of the order of 1,700,000 miles, as against 865,000 miles for our Sun.

Over the past few months Fomalhaut has been the subject of intensive study by astronomers. IRAS, the Infra-Red Astronomical Satellite which operated throughout most of 1983, had detected cool, possibly planetary material associated with the brilliant blue Vega. Other stars, such as Altair, did not give the same result; could Vega be unusual? But then, not long before IRAS came to the end of its active career, exactly the same sort of effect was found with Fomalhaut. Here, too, there was an excess of infra-red radiation, which could only come from cool material surrounding the star. This was intriguing, but it was only the beginning. Before long, the IRAS results were used to show that many stars, some of solar type, are similarly accompanied.

If the materials do really represent planetary systems, or planetary systems in the process of formation, we will have to concede a radical change in our thinking. It was once thought that the planets, including the Earth, were pulled off the Sun by the action of a passing star. If this had been true, then planetary systems would have been very rare indeed, because the stars are so widely spread in space that direct collisions, or even close encounters, can hardly ever occur (just as two gnats flying about inside the Festival Hall would be unlikely to meet head-on). Nowadays, however, it is agreed that the planets grew up by accretion from a 'solar nebula', a mass of dust and gas associated with the young Sun. What can happen to a completely normal star such as the Sun can also, presumably, happen to other stars. Vega, Fomalhaut and the other stars studied by IRAS indicate that this is really so.

If planetary systems are common, then the same may be true of life. It is not impossible that there are other civilizations in our part of the Galaxy — though it does not look as though we can hope to find them in the system of Fomalhaut.

41 STARLINK

The Royal Greenwich Observatory was founded in 1675. It was set up in the Park, but during the 1950s the working instruments were shifted to the clearer skies of Herstmonceux, near Hailsham in Sussex. (From Herstmonceux come the famous 'six pips' time-signals, first broadcast at 9:30 pm on February 5 1924, so that they have a history stretching back for over 60 years, and the Royal Observatory is still regarded as the time-keeping centre of the world.) Today the largest telescope, the Isaac Newton Reflector, has been given a new mirror and moved to La Palma in the Canary Islands, where it will be joined shortly by the even larger William Herschel Telescope. It is probably in the field of instrumental development that Herstmonceux itself is now most important. To quote the Director, Professor Alec Boksenberg: 'It is the instrumentation used with a large telescope which brings in the scientific results. Astronomers no longer look through a telescope by eye and jot things down in a notebook; that is hopelessly out of date. Nowadays, discoveries are made with highly sophisticated instrumentation. We have to be adaptable and to respond quickly to new advances — and also initiate new advances; we must continually provide new recording devices to exploit the great new telescopes to full advantage.'

Britain has always been very much to the forefront. It is fair to say that some of the most important instruments of electronic design were developed here, and Herstmonceux retains its pre-eminent position. There is also 'Starlink' which is purely British and which is the envy of astronomers in other countries.

Basically, Starlink provides a means of analyzing observations and combining them with results obtained elsewhere. The system consists of a set of eight identical computers situated at major astronomical centres in Britain (at the Royal Observatory Edinburgh, for instance). The computers are exceptionally fast, and can handle large blocks of data. A single night's observing on a telescope such as the INT can provide enough material to keep an 'old-fashioned' astronomer busy for a year at least, but with the computers the whole programme can be played back and analyzed in detail. The raw observations can also be amplified, by the use of various filters and other devices.

For example, consider two galaxies which have been under study by Dr Robert Fosbury and Dr Jasper Wall, at Herstmonceux. They suspected that one galaxy, a giant elliptical system, was distorting its neighbour, a spiral. Observations analyzed with the Starlink system have confirmed that this is so. The spiral is being torn apart, and in something like 100 million years the beautiful spiral will have been destroyed and swallowed up by the elliptical system. Another case shows two spirals which are colliding, producing a mélée which is incredibly complex. Again, the raw data had to be processed and analyzed with the new techniques before the collision could be proved. And remember, a galaxy such as our own contains about a 100,000 million stars.

Starlink has come to stay. It is only the beginning of a new story which is certain to lead us on to a better understanding of the universe in which we live.

Impression of the forthcoming Hubble Space Telescope.

42 THE VIOLENT WORLD OF VENUS

Venus, the lovely 'Evening Star', is the closest of the planets; at its nearest to the Earth it may approach us to within 25 million miles, so that it is then only about 100 times as far away as the Moon. In size and mass it is almost the Earth's twin. Before the Space Age we knew very little about it, because no telescope could see through its dense, cloudy atmosphere; it was even thought possible that primitive life-forms might exist there. Today we know better.

Venus is closer to the Sun than we are, moving at a distance of only 67 million miles in an almost perfectly circular orbit. Its 'year' is equal to 24¾ Earth-days. Its own 'day' is surprisingly long; Venus takes 243 Earth-days to make one turn, and to make matters even more curious it spins from east to west, so that if an observer on the planet's surface could see the sky the Sun would appear to rise in the west and set in the east. However, the sky would never be visible from Venus. The planet is enveloped in a dense atmosphere made up chiefly of the heavy, unbreathable gas carbon dioxide. Because carbon dioxide has what is termed a greenhouse effect, and shuts in the Sun's heat, it was natural to assume that the surface temperature must be high, but before the flight of the first successful planetary probe, Mariner 2, in 1962, that was about the sum total of our knowledge.

Since then, space-craft sent up by both the Russians and the Americans have led to a complete change in outlook. The surface temperature is not far short of 550°C which is much too hot for any liquid water to exist. The atmospheric

Overleaf *Impression of the Venera probe on the surface of Venus by a Russian artist.* **Inset** *The surface of Venus, from Venera 9. Part of the spacecraft is shown, together with the rock-strewn landscape.*

Right *Clouds over the night side of Venus, photographed in infra-red by D. A. Allen at Siding Spring Observatory, Australia.*

Below *Three views of Venus from Mariner 10. The arrows show the same cloud structure, and give a rotation period of four days (retrograde) for the upper clouds.*

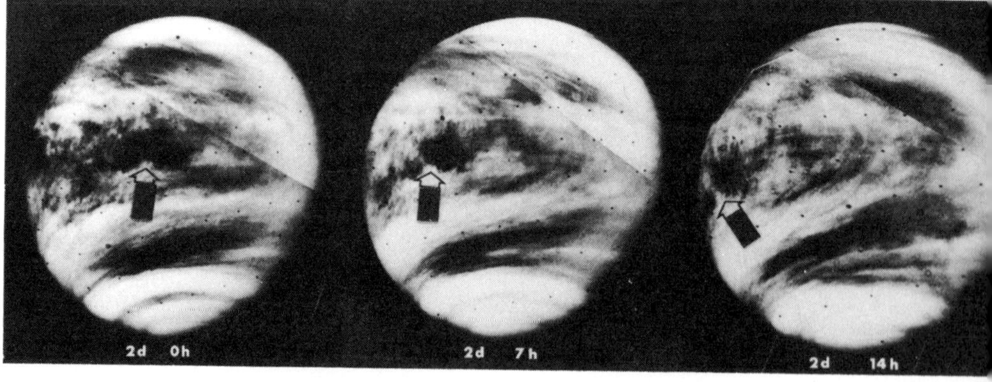

pressure at the surface of the planet is over 90 times that of our own air at sea-level. And surprisingly, the upper clouds have a rotation period of only 4 days, even though Venus itself spins so slowly.

The lowest part of the atmosphere has been described as superheated, corrosive smog. Above this layer come the clouds, which encircle the whole planet. The winds are reasonably strong, and the clouds themselves are not like ours; they contain large quantities of sulphuric acid. As the acid droplets condense, they fall to the ground, but are evaporated before they reach the bottom of the cloud-layer, which is thought to be about 30 miles above the surface — so that the deadly 'rain' is confined to the clouds.

In 1983 D. A. Allen and J. Crawford, using the Anglo-Australian Telescope at Siding Spring in Australia, studied Venus at infra-red wavelengths, and recorded details on the night side. These details seem to be clouds, shown up by infra-red solar radiation scattered through the daylit atmosphere to the dark hemisphere. Their rotation period seems to be 5.4 days, so that they lie lower down than the visible clouds; Allen and Crawford conclude that they are inside the main cloud-deck, about 30 to 40 miles above the actual surface.

The Soviet Venera probes, which made controlled landings on the surface, have sent back pictures showing a gloomy, rock-strewn landscape under a lurid orange sky. The surface has also been mapped by radar, both by Soviet orbiters and by the US Pioneer which began circling the planet in 1978. On Venus there is an extensive rolling plain, together with two upland areas, now picturesquely named Ishtar and Aphrodite. Ishtar, the loftier of the two, is about the size of

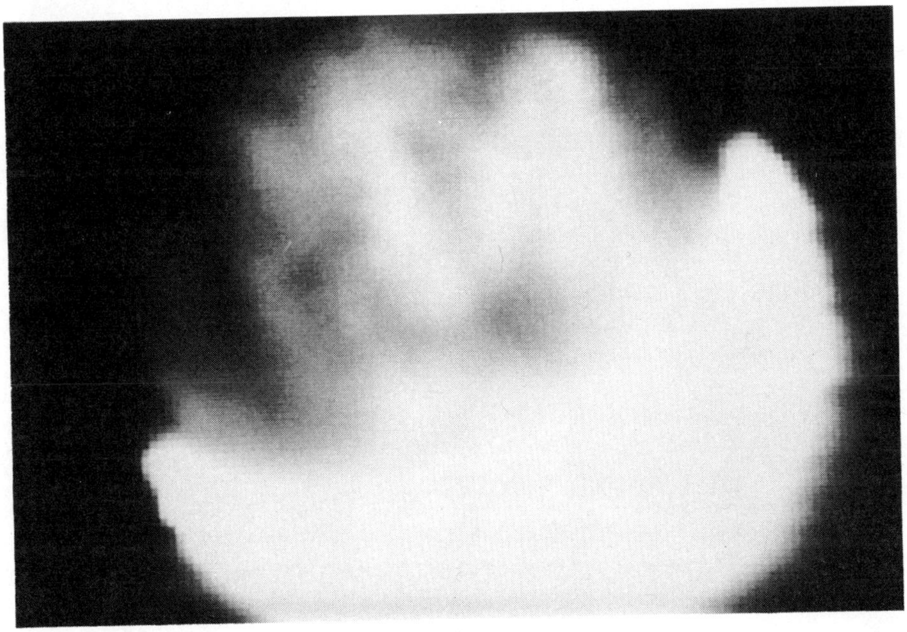

Australia, and contains the highest peaks on the surface: the Maxwell Mountains, which rise to 35,000 ft (though altitude measurements are bound to be somewhat arbitrary, because there is no 'sea level' on a waterless world). Adjoining Aphrodite, the second upland area, there is a huge rift valley much larger and deeper than anything on Earth, and comparable with the Valles Marineris on Mars. There are also many craters, presumably of volcanic origin. Beta Regio, another lofty area, contains two features, Rhea Mons and Theia Mons, which are almost certainly shield volcanoes, similar in type to the Hawaiian volcanoes on Earth but considerably larger and more massive. Atla Regio, adjoining Aphrodite, is also volcanic in nature.

Are the volcanoes of Venus active now? Some of them, notably the Maxwell Mountains, are believed to be either dormant or extinct, but there is strong evidence of activity in both Beta Regio and Atla. Lightning occurs there, and lightning is often associated with vulcanism. Moreover, it seems that the amount of sulphur dioxide in the atmosphere increased dramatically in 1978, since when the level has been falling again. Dr Larry Esposito, of the Universty of Colorado, interprets this as being due to a tremendous volcanic outburst, beside which the Earth's Krakatoa eruption of 1883 would seem very mild. 'The natural way to explain this,' he writes, 'is to assume that sulphur dioxide came up from the surface and then dropped in the form of acid rain, indicating a volcanic eruption shortly before then.'

One important difference between Earth and Venus is that 'plate tectonics' do not apply to Venus in the same way. The Earth's crust is divided into several large plates which drift very slowly around; volcanic activity occurs at the plate boundaries and over certain definite regions. A good example is provided by the volcanoes in Hawaii. Of the two main volcanoes on Big Island, Mauna Loa is active while Mauna Kea is not. A volcano lies above a 'hot spot' or plume deep inside the Earth's mantle. Mauna Kea used to lie above this plume, but it has now moved away, and all activity has ceased; Mauna Loa is now over the plume, and will remain active until it too drifts away and becomes extinct. But on Venus there is no such movement, so that a volcano above a plume remains there for a very long time indeed. This is probably the case with Rhea and Theia, and with the volcanoes in Atla. If so, then new eruptions may occur at any time.

With its dense, corrosive atmosphere and its searing heat, Venus was known to be hostile enough; the probability of violent eruptions makes it even worse, and the chances of any manned expeditions there, at least in the foreseeable future, seem to be nil. With the naked eye Venus looks calm and peaceful; it requires a considerable effort of the imagination to realise that conditions below those shining clouds approximate closely to the conventional idea of hell!

43 STAR BIRTHS AND STAR DEATHS

How are the stars born? How do they evolve, and how do they die? Until recently these were questions which could not be answered with any degree of

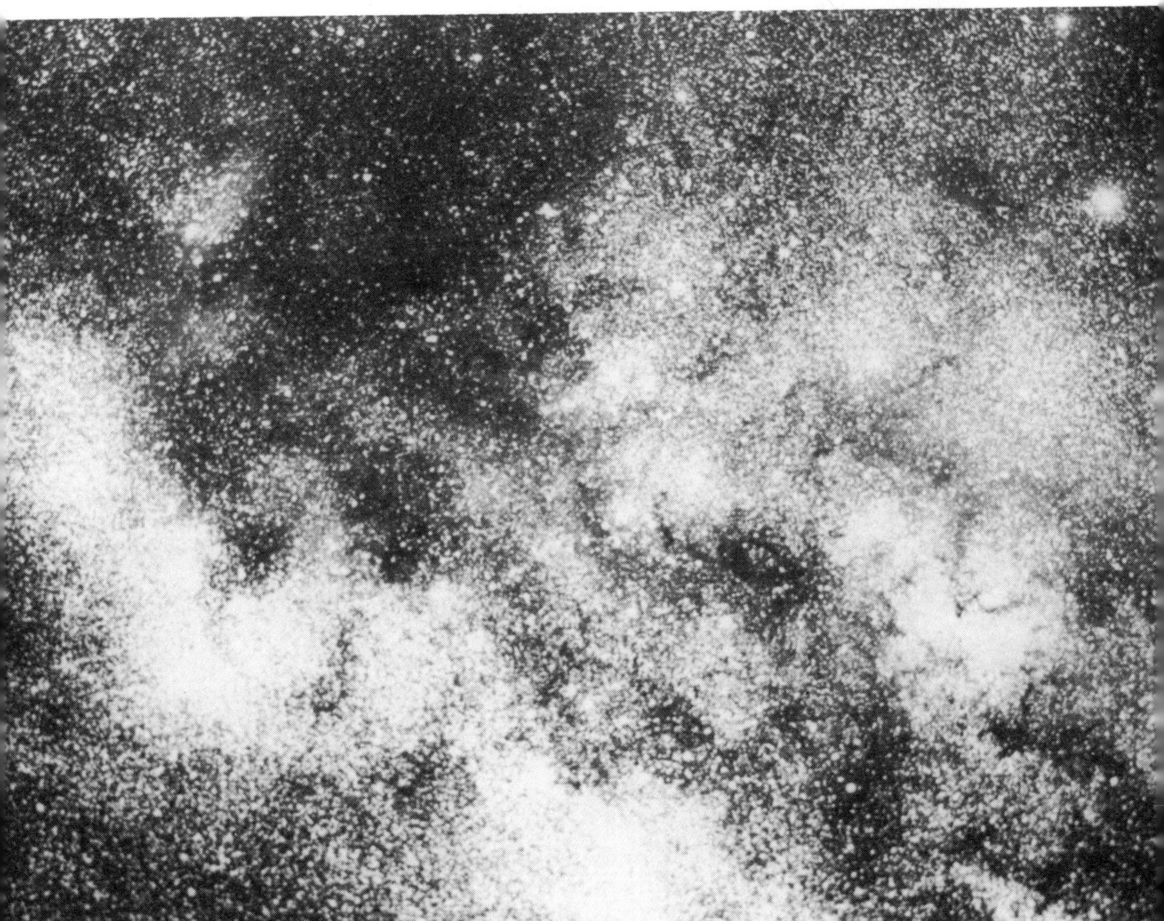

The Milky Way in the region of Cepheus and Lacerta. The bright star to the upper right is Alpha Cephei.

certainty, but today we have at least been able to draw up an accurate general picture of the lifetime of a typical star.

There are two main difficulties. First, only the Sun is close enough to be examined in any detail. The other stars are so remote that they look like nothing more than points of light, and we have to depend upon instruments such as the spectroscope. Secondly, the lifetimes of the stars are very long by everyday standards. Our Sun must be older than the Earth — and the age of the Earth can be fixed fairly precisely at between 4,600 million and 4,700 million years. This means that we cannot see a normal star evolving. All we can do is to study stars

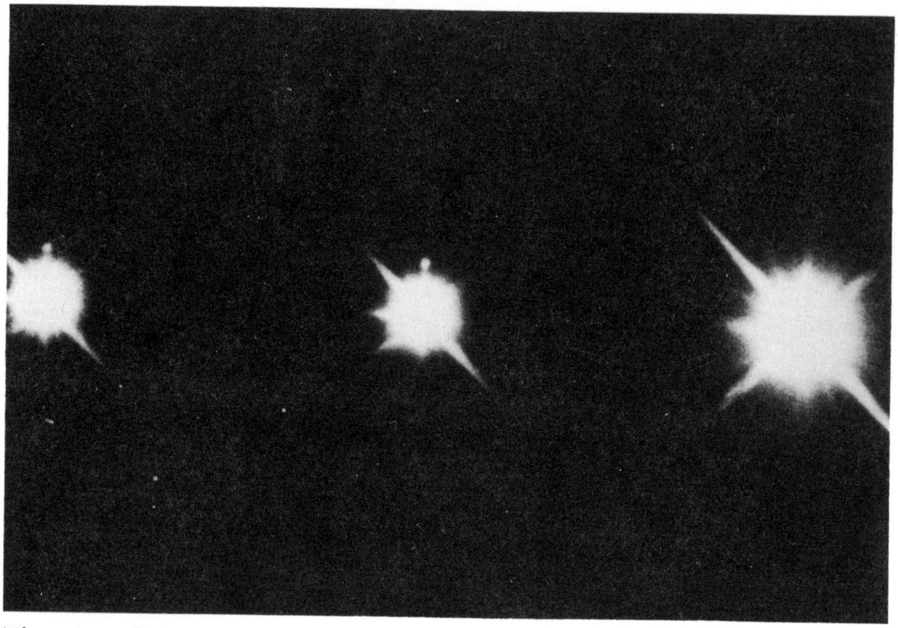

Three views of Sirius. The 'Pup' appears as a tiny speck.

which are in different stages of their careers, and see how we can piece the story together.

A star begins inside a nebula, made of dust and gas. Many nebulæ are known, and some, such as Messier 42 in Orion, are easily visible with the naked eye. Condensations inside a nebula produce stars — not single stars, but whole clusters. When the cores of these fledgling stars become sufficiently hot and dense, energy production begins and the star starts to shine. As the output of energy increases, the material which has not been 'used up' to form the original cluster is blown away and eventually only the cluster itself remains. Open clusters are common enough and again there are naked-eye examples, of which the most famous is the cluster of the Pleiades or Seven Sisters, a prominent feature of the night sky for several months in each year. The Pleiades cluster is probably only about 60 million years old, and some nebulosity remains. A more evolved cluster is Præsepe or the 'Beehive' in Cancer, which also is visible with the naked eye, lying about midway between Regulus in Leo, the Lion and the Twins, Castor and Pollux. Præsepe is much more scattered than the Pleiades, with no nebulosity left. Its age is thought to be about 400 million years.

Gradually, the stars in a cluster begin to disperse because of the gravitational effects of other stars which do not belong to the group. However, in many cases two or more stars remain associated with each other, making up binary systems such as Mizar in the Great Bear — or even more complex systems such as Castor, where there are two bright, white pairs of stars and a third pair made up of two dim Red Dwarfs. The view from a planet in the Castor group would indeed be spectacular.

Binary stars are plentiful; indeed, it may well be that single stars such as the Sun are in the minority. As time passes, the original open cluster becomes more and more scattered, until it is difficult to identify at all; it has become what is usually called a 'stellar association'. Again the Great Bear provides a good example. Of its seven main stars, five are members of the same association; they are moving through space in the same direction at the same rate. The other two stars in the pattern, Alkaid and Dubhe, are moving in the opposite direction, so that over a sufficiently long period of time the familiar Great Bear will become distorted and finally unrecognizable.

As we have seen, a normal star of solar type is producing its energy by nuclear reactions taking place deep inside it. When the 'fuel' runs low, the star has to rearrange itself. The core shrinks and becomes hotter, while the outer layers billow out and cool down, so that the star becomes a Red Giant. This will happen to the Sun in perhaps 5,000 million years from now, and the Earth cannot hope to survive, at least as a habitable globe.

The Veil Nebula in Cygnus — undoubtedly an ancient supernova remnant — photographed with the Palomar 200 in reflector.

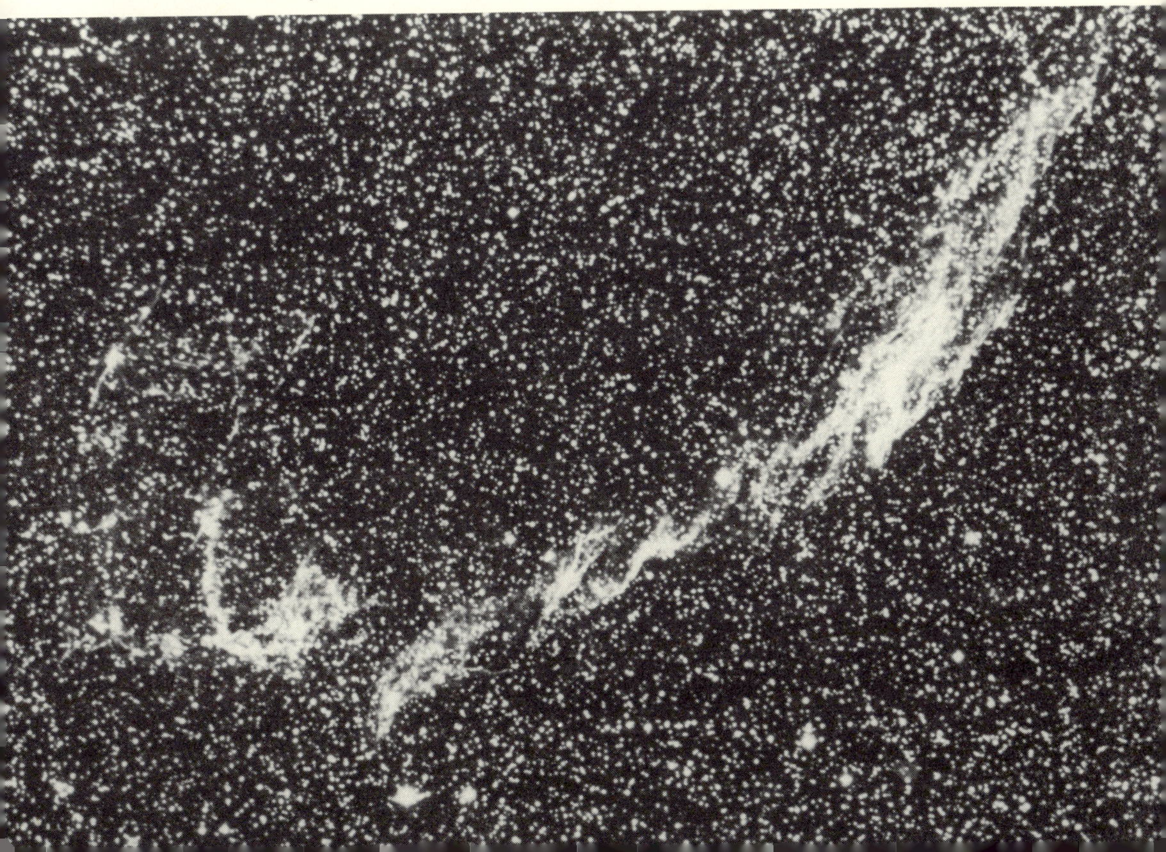

The outer layers of a Red Giant are unstable; they swell and shrink alternately. Eventually they are thrown off altogether — either in one event or in a series of events; at any rate, the process is a comparatively gentle one. The thrown-off atmosphere produces a glowing shell which we see as a planetary nebula (not an apt name, because we are not dealing with a true nebula, and certainly not with a planet). Messier 57, between the naked-eye stars Beta and Gamma Lyræ near the brilliant blue Vega, is a classic planetary nebula; a moderate-sized telescope will even show the central star which is simply the core of the old Red Giant.

The shell of a planetary nebula expands steadily, and in a fairly short time by cosmical standards it will cease to glow. We are then left solely with the collapsed core of the old star, which is extremely dense; it has become a White Dwarf, and is 'bankrupt', so that it can produce no more energy by nuclear reactions. Sirius, the brightest star in the sky, has a White Dwarf companion which is as massive as the Sun but is only 26,000 miles in diameter — and a mere one ten thousandth as luminous as Sirius itself.

With a solar-type star this marks the onset of the last phase. Gradually the White Dwarf cools down, and ends its career as a cold, dead globe. More massive stars end their careers dramatically by exploding as supernovæ. And with an even more massive star, the end product may be a Black Hole, from which no radiation of any kind can escape.

Such is the overall picture — but one fact is all-important. Stars expel material; supernovæ explode and distribute their mass into space. The material is then used to make new generations of stars, and the cycle is repeated over and over again. Our Sun is not a first-generation star, which means that the material making it up was once contained inside an earlier star. This is also true of the entire Solar System, including the Earth. It is rather sobering to realise that the atoms making up you and me were once part of a star which died thousands of millions of years ago!

44 WHEN THE SUN FLARES

On February 14 1980 an important unmanned artificial satellite was launched. Its official title was the Solar Maximum Mission, though it is almost always known simply as Solar Max. As the name implies, it was designed to carry out studies of the Sun, and there was a special reason for dispatching it at this particular time. The Sun may be regarded as a variable star; every 11 years or so it is very active, with numerous groups of sunspots and violent outbreaks known as flares, after which the activity dies down before starting to build up again toward the next maximum.

Maximum occured in 1980. Therefore, for the first time, astronomers could have a whole array of instruments operating in space to monitor the Sun at its most violent. Many investigations cannot be properly carried out from the

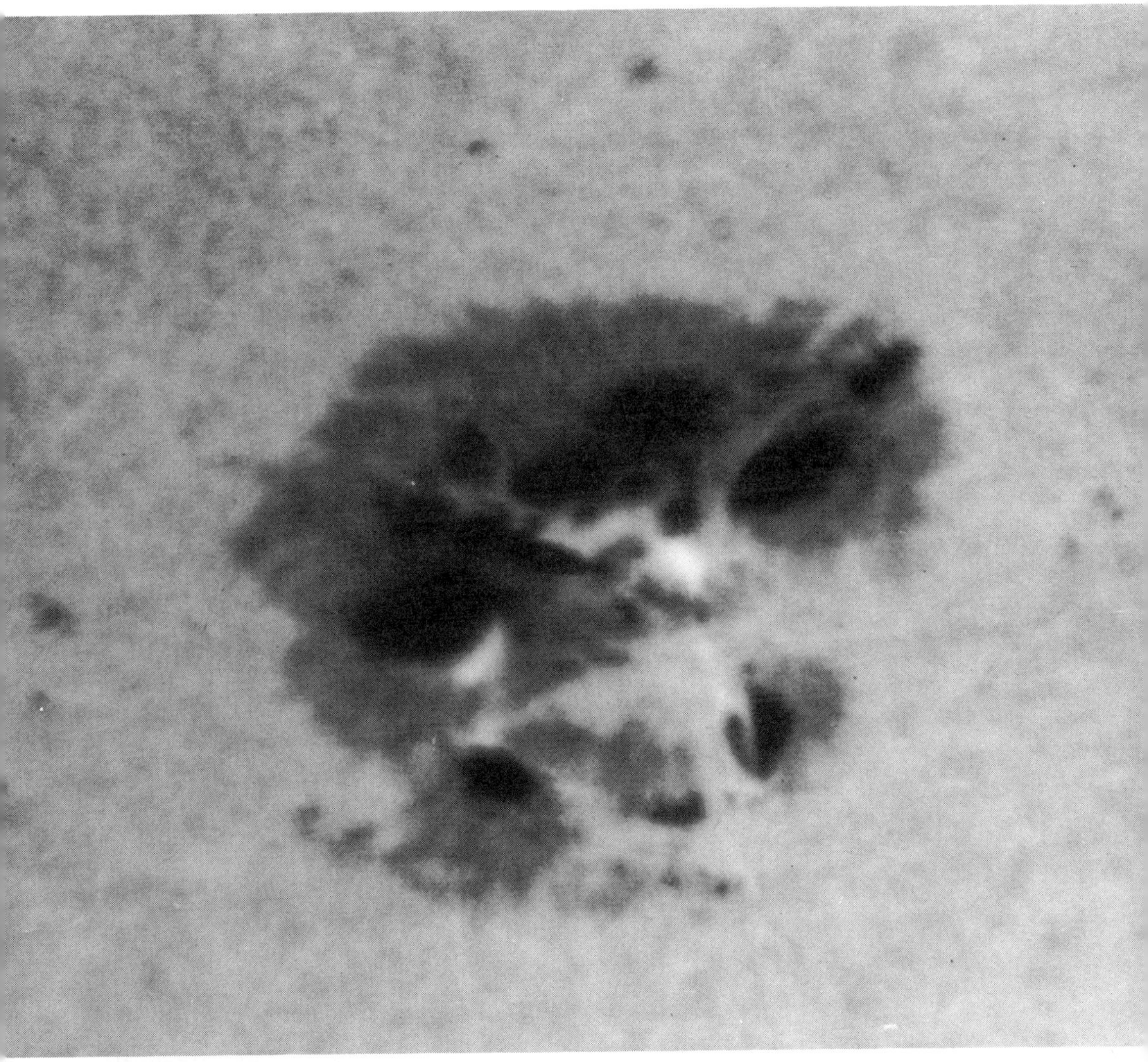

Solar flare, August 7 1972.

Earth's surface; for instance, the flares on the Sun send out short-wave radiations (X-rays in particular) which are blocked out by the protective screen of our upper atmosphere.

For nine months all went well, but then Solar Max developed a serious fault. It lost its 'attitude control', so that it could no longer be aimed in the required direction, and a great deal of the most valuable information was lost.

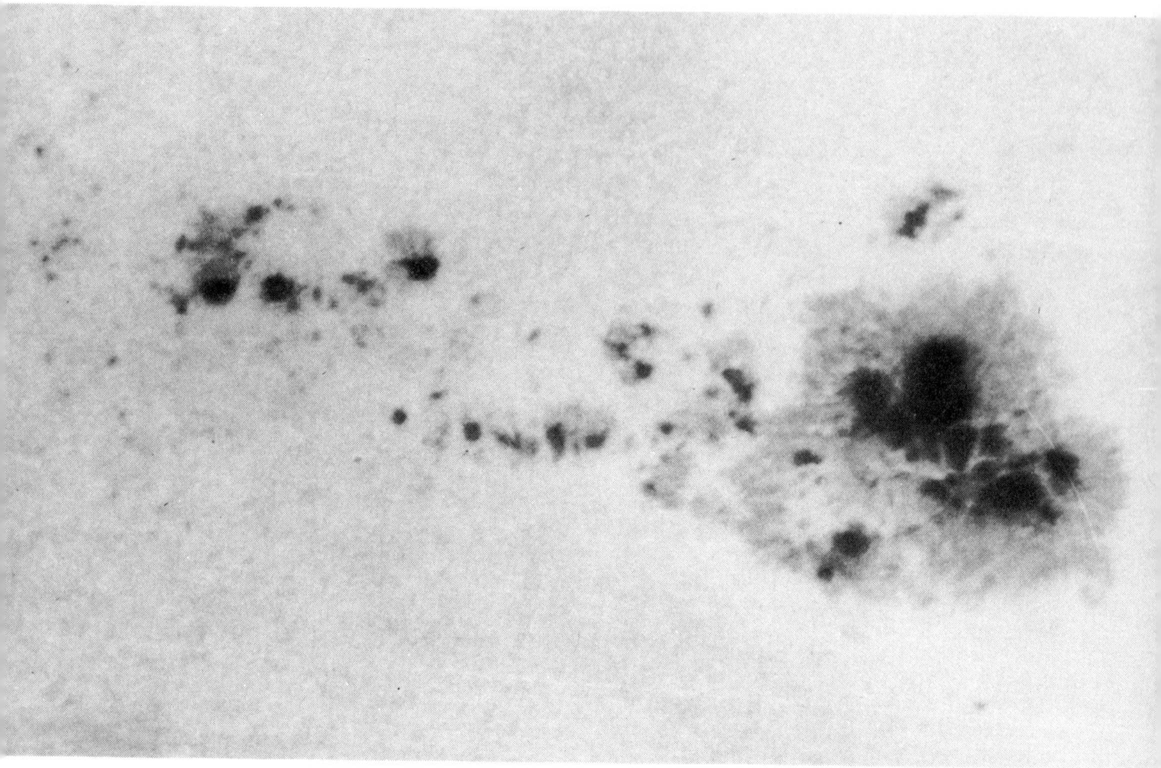

Large sunspots.

What could be done? Solar Max was orbiting at a height of approximately 300 miles above the ground — and this is within the range of the Space Shuttle. An ambitious plan was worked out. A Shuttle crew would go up to Solar Max, bring it into the space-craft's bay, repair it and re-launch it. This was precisely what was done in the spring of 1984. The names of the five astronauts deserve to go on record: Robert L. Crippen, Francis R. Scobie, George D. Nelson, James D. van Hoften and Terry J. Hart.

Rendezvous was achieved easily enough, but then the problems mounted. Solar Max was spinning, and the first need was to stop this spin. Dr Nelson made a 'free flight' to the Shuttle and did his best to latch on to the recalcitrant satellite, but only made things worse. The outlook seemed bleak; but then an attempt to grab Solar Max by an extended arm was successful. The satellite was edged into the bay, and duly repaired — not without difficulty; one of the astronauts commented that it was rather like trying to perform a delicate surgical operation while wearing boxing-gloves. But in the end all went well, and Solar Max, now fully operational again, was returned to space. With luck it

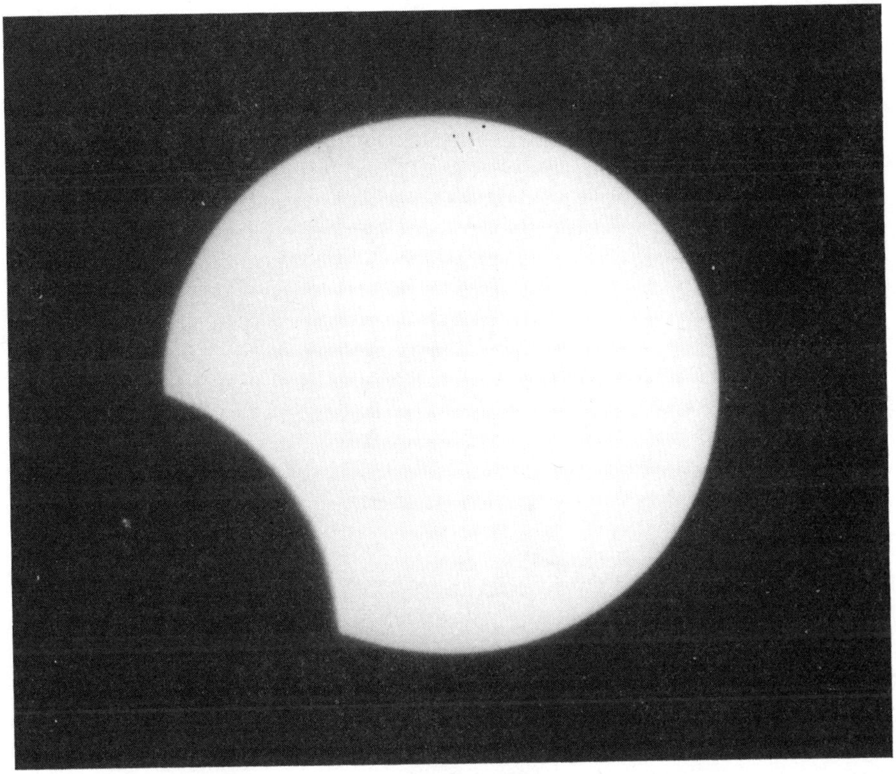

Partial eclipse of the Sun.

will continue to operate all through the coming minimum of solar activity, and should still be functioning at the time of the next maximum, which is expected to be around 1990–1991.

The Sun is a normal star; the temperature of its bright surface or photosphere, where sunspots occur, is nearly 6,000°C. Above the photosphere comes the 'colour sphere' or chromosphere, not visible with the naked eye except when the Moon covers the Sun during a total solar eclipse — though instruments based on the principle of the spectroscope make it possible to study the chromosphere at any time, as well as the huge masses of glowing hydrogen which were once (misleadingly) known as Red Flames and are now called prominences. Beyond this region comes the corona, the Sun's outer 'atmosphere' which again is visible with the naked eye only during a total eclipse. At times of spot-minimum the corona is relatively restricted; when the Sun is at its most active the coronal streamers stretch across the darkened sky. It is a pity that the next total eclipse visible in England will not occur until August 11 1999.

One curious fact is that the temperatures in the chromosphere, prominences and corona are very high, rising to well over a million and a half degrees in the corona. Yet this does not indicate any real 'heat'. Scientifically, temperature is defined by the rate at which the various atoms and molecules move around. In the corona these speeds are very high but there are very few particles, relatively speaking, because the corona is so rarefied. For many years the reason for the high coronal temperature was unclear. It was suggested that the immense disturbances lower down generated tremendous sound waves which would produce the high temperatures — in fact, Titanic 'solar sonic booms' — but space research methods have shown that there are no shock-waves at the base of the corona, so that the answer must lie elsewhere. In all probability the root cause is magnetic.

Flares are brilliant, short-lived outbreaks, usually over sunspot groups. Very occasionally they are brilliant enough to be seen in ordinary light (the first flare to be seen directly was that of September 1 1859, by the British astronomer R. C. Carrington) but most are invisible without spectroscopic methods, when the Sun can be examined in the light of hydrogen only. A flare mechanism has been likened to a lightning strike, though on a far larger scale. There is a sudden release of the energy which has been stored in the magnetic fields of an active region; as this energy is released, in the manner of a snapping spring, there is an outbust which would dwarf more than a million hydrogen bombs. Thousands of millions of tons of material are ejected, and the flares emit charged particles as well as short-wave radiations. The X-radiations and ultra-violet reach the Earth in little over eight minutes; the particles take from one to four days, producing magnetic disturbances, interference with radio transmissions, and also the lovely displays of polar lights or auroræ.

We cannot yet pretend that we know the full details of why sunspots break out; neither do we really understand the flares, though it seems that a flare is produced when two loops of magnetic fields collide and interact, so that the high-energy charged particles flash through the loops and make their presence known. Flares generally erupt several thousands of miles above the Sun's bright surface, in the lower corona or in the chromosphere; they last for only a few minutes, but inside them the temperature may soar to as much as ten million degrees.

Obviously we want to learn all we can about the Sun; in particular, we want to learn what effect the solar cycle has upon the Earth's climate and weather. Already Solar Max has provided important new information, notably with regard to the 'solar constant' — that is to say, the total amount of energy which the Sun is emitting. There are unquestionably slight fluctuations, though we do not yet know what may happen over a longer period of time.

Further satellites are being planned. For instance there will be the Solar Optical Telescope, with a 48 in mirror, designed to look at the Sun in ordinary light under conditions of perfect seeing. Then there is the International Solar Polar Mission, scheduled for the coming maximum; the probe will climb far above the plane of the Earth's orbit round the Sun, and study regions of the solar globe about which we know little as yet. In the meantime Solar Max will, it is

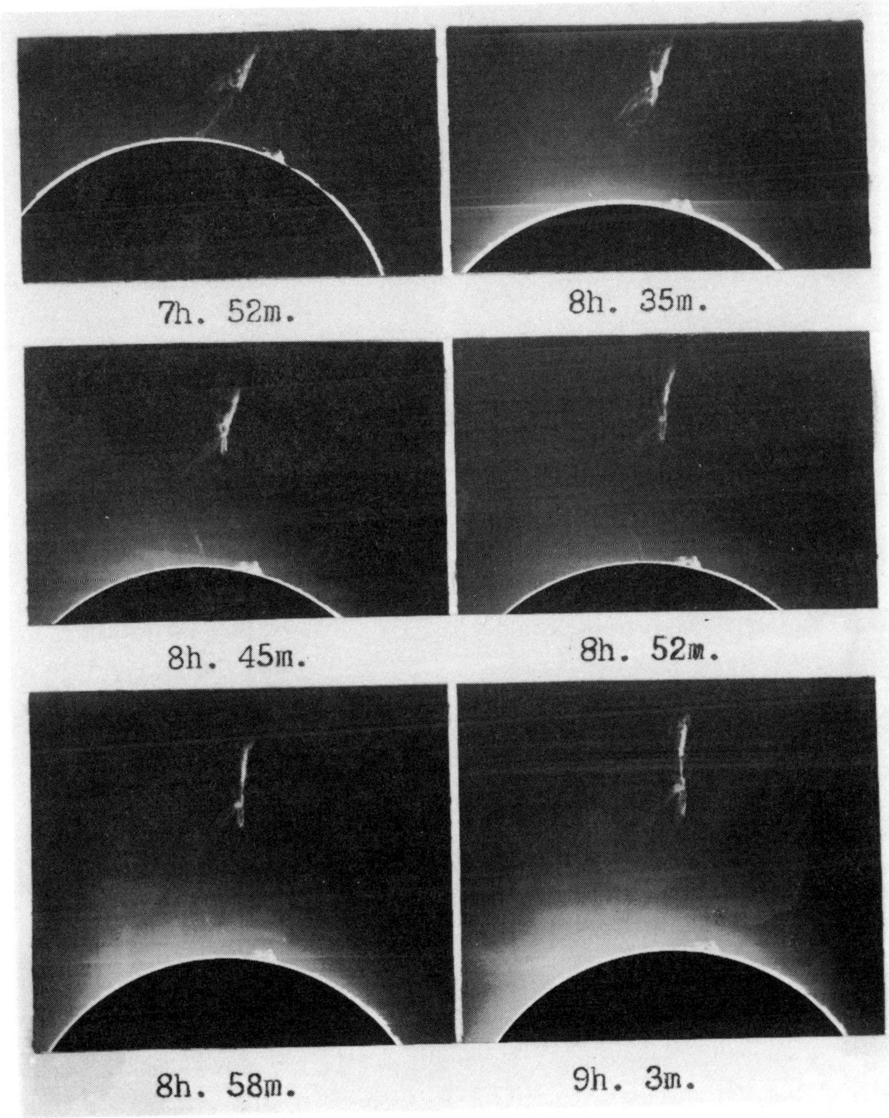

Eruptive prominence, November 19 1928. Kodaikanal. Maximum height 567,000 miles.

hoped, fill in some of the missing links in our understanding of the Sun and its magnetism. At any rate, there can be no doubt that the rescue mission has been very worth while, and represents one of the greatest triumphs in space research date.

45 BUILDING UP THE UNIVERSE

The question of the origin of the universe has been very much in astronomers' minds since the very early years of true science. Unfortunately, though we have become knowledgeable so far as details are concerned, we are still woefully weak in fundamentals. The material making up the galaxies, the stars, the Sun, the Earth, you and me must have come from somewhere or other, but we have no idea of the nature of the 'creation' itself.

It is generally believed that the universe began at a set moment in time, between 15,000 million and 20,000 million years ago, with a preference for the lower figure. There was what may be called a 'big bang' when space and time came into existence. During its first few seconds the universe was extremely hot and extremely small; by now it has expanded, so that we can see out to at least 13,000 million light-years. All that remains of the 'big bang' is weak radiation, indicating a temperature of 3° above absolute zero, permeating all space. Starting from this assumption we can even work out the ways in which the various chemical elments were formed. This may not bring us any closer to the fundamental creation, but at least it is better than nothing.

The main agents in the build-up of the universe as we know it today are the stars, which, as we have seen, are nothing more nor less than huge nuclear reactors. Hydrogen is the basic 'fuel', and in normal stars such as the Sun it is being converted into helium. This has been known for some time; the principles were worked out just before the war, mainly by Carl von Weizsäcker in Germany and George Gamow in America. It is also possible to calculate how much helium has been 'cooked' inside stars since they were formed, and average this over all the matter contained in the galaxies. The result is about four to five per cent of the material in the universe. And here we come to our first difficulty; the observed amount of helium is nearer 30 per cent. Something is wrong. If the observed quantity of helium cannot have been produced inside stars, we come back to the idea that it must have been formed at the time of the original big bang itself.

The first theorist to come to this conclusion was George Gamow in 1940. Unfortunately, his ideas were not taken very seriously at the time. This was partly due to the personality of Gamow himself. He was Russian, and spent some time in Denmark, Germany and England before going to America, where he spent most of the rest of his life. There was never any doubt about his brilliance, but some of his daring ideas were treated with scepticism, and certainly he was unconventional. One anecdote about him will show what is meant. He attended an important conference about star clusters, and slept throughout almost the whole of the proceedings — not quietly; his snores reverberated across the conference hall. Yet a few days later the organiser of the meeting received a telegram from Gamow which showed that, by a brilliant flash of insight, he had found a way to measure the ages of the clusters. The most brilliant stars are bluish and short lived; white stars are less brilliant and have longer lives; yellow stars and red stars are more leisurely still. Therefore, a cluster in which the leading stars are blue must be young, while a cluster in which the chief members are red must be old — the blue stars have already disappeared. This is now the classic way of measuring the ages of clusters. (One colleague commented, rather acidly, that Gamow half-asleep could think more clearly than any other astronomer wide awake.) But such eccentricities did not endear him to all his contemporaries.

Gamow believed that at the time of the original big bang, the universe was almost incredibly hot, and that the conversion of hydrogen into helium was very rapid, explaining the 30 per cent now found. Gamow went on to suggest that by using this model he could explain the building-up of all the elements, starting from hydrogen and working through to the heaviest elements known to occur in nature. Finally, he predicted the existence of the weak background radiation. He was correct about the background, even though it was not actually detected until the 1960s. But even Gamow, with his intuition, was wrong in his belief that his 'hot big bang' model could explain the whole sequence of elements. The true picture must be more complicated than that.

There were, of course, attempts by other theorists to do away with the big bang altogether. In 1948 a group at Cambridge, headed by H. Bondi and T. Gold (later joined by Sir Fred Hoyle) put forward the 'steady-state' idea, according to which the universe has always existed, and will exist forever; as old galaxies disappear, new ones are formed from material which is created spontaneously out of nothingness. This led on to a new attempt to explain the existence of heavy elements. However, the theory failed to stand up to observational tests. It would mean that the average number of galaxies per unit area of the universe would always be much the same as it is now; if we could come back in, say,a million million years' time we would see the same numbers of galaxies as we do at present — though they would not be the same galaxies. We cannot travel in time, in the manner of Dr Who, but we can 'look backward' in time, because when we observe galaxies thousands of millions of light-years away we are seeing them as they used to be thousands of millions of years ago; and the distribution in those remote parts of the universe differs from that in our own region, so that the universe is not in a steady state, and the theory is wrong.

There was also the problem of the substance called deuterium, which is a heavy form of hydrogen. Stars do not create deuterium; they destroy it — and so on the picture presented by the Cambridge group, there should be little deuterium around. One eminent astronomer, Willy Fowler, said in the 1960s that if deuterium were ever found outside the Earth, he would eat his hat. Presumably he did so, because since then deuterium has been abundantly detected in the space between the stars. Combined with the unexpectedly high percentage of helium, this means that element-building must have occurred in places other than the stars; and this seems to bring us back to the big bang.

Efforts to work out other pictures of the history of the universe are still being made. One of them abolishes the big bang, replacing it with a set of super-massive stars which would have had to exist before the formation of the galaxies themselves. We also have the decidedly confusing 'inflationary' models, in which there are various big bangs, each of which produces its own particular set of circumstances. But always we come back to the basic problem — what created the material in the first place? One day, perhaps, we may find out, but we certainly do not know yet.

46 A NIGHT AT LOS MUCHACHOS

In April 1984 I was privileged to take part in a commissioning run with the Isaac Newton Telescope, the 100 in reflector which has been installed at the Los Muchachos Observatory on La Palma, in the Canary Islands. The INT was transferred there from its old home at Herstmonceux in Sussex, and given a new mirror; as well as being one of the world's larger telescopes it is the most modern, and is completely computerised.

The INT is impressive by any standards, and from its site, at an altitude of nearly 8,000 ft, the observing conditions are probably as good as anywhere in the world — even the top of Mauna Kea.

A commissioning run is really a test of the telescope, to see how it is behaving and what adjustments have to be made. On this particular night our main target was the Ring Nebula in Lyra, which is about 1,400 light-years away and is a comparatively faint object. Small telescopes show it as a dim smudge of light midway between the naked-eye stars Beta and Gamma Lyræ, but large instruments are needed to bring out its details. Our aim was to secure a colour video picture of it, something which had not previously been attempted for an object outside the Solar System. This was also the first occasion when a telescope of such size had been used in a television programme, which made it doubly memorable.

The night was clear; the sky was dark and cloudless, and as soon as possible Dr Paul Murdin, who is in charge of the British telescopes on La Palma, supervised the opening of the dome and the preparation of the INT. We settled ourselves in the control room (there was nobody actually inside the main dome itself) and as soon as the co-ordinates were set, the telescope swung automatically to the correct position. The Ring Nebula came into the view-field of the small finder telescope, and as soon as it had been brought to the centre we switched to the INT itself.

The Ring is a planetary nebula: that is to say, a late stage in the evolution of a normal star. The remnant of the star can still be seen in the centre of the shell of gas; it is faint because it is comparatively small, but it is extremely dense, and its surface temperature is very high indeed — perhaps as much as 100,000°C against a mere 6,000°C for our Sun. Nuclear reactions have stopped, and the star will eventually fade away to the Black Dwarf condition. At present the Ring Nebula is about half a light-year in diameter, and is expanding at a rate of 12 miles per second, indicating that it began its planetary nebula phase no more than about 20,000 years ago.

Dome of the INT at La Palma, as I photographed it in 1981.

As we watched the television screen, the Ring came sharply into focus, giving a clear black-and-white picture; the structure was plain enough, and in addition to the central star there were two others, though we knew that they lay in the foreground and were not actually associated with the Ring. Now it was time to use colour filters: first blue, then green and finally red. Each gave a perceptibly different picture, to bring out different aspects of the Nebula. For example, the central hot star is bluish, and showed up much more strongly with the blue filter than with the red; for one of the foreground stars the reverse was true, showing that it had a much lower surface temperature.

Next the three pictures were combined — overlaid, in fact — so as to produce what was to all intents and purposes a colour video. The effect was remarkable. Certainly I had never seen anything like it; it was not only beautiful, but also breath-taking. All the intricate details in the Ring were enhanced, and it was difficult to believe that we were looking at an object well over a thousand light-years away.

Was it a true colour picture? Not completely — just as a television view of a cricket pitch will enhance the greenness of the grass and the whiteness of the players' flannels. But if one could really go close to the Ring, it would look very much like the image as we saw it from the control room of the INT.

Isaac Newton Telescope (Royal Greenwich Observatory).

Lovely though they are, pictures of this sort are of limited scientific value. To obtain the most useful results, the light is sent into a spectrograph and is split up. This makes it possible to identify the elements making up the object, as well as its temperature and other details. We were able to manage this, too, and it was clear that the telescope was working really well. Later in the night we turned to other objects — the superb Whirlpool Galaxy in the constellation of the Hunting Dogs, which is a face-on spiral 35 million light-years away, and then the huge elliptical system Messier 87 in Virgo, with its curious 'jet' which indicates that it is extremely active, possibly with a Black Hole at its centre. But impressive though they were, it is that first view of the Ring which I will always carry in my mind.

By now the INT has begun its main programme of research, and it will be used on every clear night. A smaller reflector, the 40 in Kapteyn Telescope, is also in operation, and before long there will be others too. A new 100 in reflector is being planned as a combined project by Norway, Sweden, Denmark and Finland. Meanwhile, work is continuing on the even larger William Herschel reflector, which has a mirror 165 in across. The telescope is being made by the firm of Grubb Parsons in Newcastle (sadly, their last mirror before closing down that part of their works) and the construction of the dome for it is well under way at a site some way below the volcano's summit, where the Rocks of the Boys stand. It is hoped that the WHT will be in operation before the end of 1986.

I came away from La Palma full of hope for the future. There can be little doubt that many important discoveries will be made at the Observatory in the years to come, and it is good to know that in so tremendous a project Britain is playing the leading role.

47 MISSIONS TO COMETS

Comets are very much in the news at the present moment. This is understandable, because the most famous of all comets, Halley's, is on its way back, making its first return to the Sun since 1910. This will be an event of tremendous scientific importance. At the last return there were good photographic results in black-and-white; but the type of the equipment in use today lay far in the future, and the idea of sending a rocket probe to rendezvous with the comet, or even pass right through it, would have been dismissed as ridiculous. Yet this is precisely what is going to happen in March 1986, provided that all goes well.

Our knowledge of comets is much less than might be expected, and it has even been said that we know more about the core of the Sun than we do about the nucleus of a comet. We have never seen either. Comets are awkward things; as soon as a comet comes close enough to the Earth and the Sun to be properly studied, the ices in its nucleus start to evaporate, and the heart of the comet is hidden. This is why comet probes are so urgently needed.

Of course, not all comets have tails; many of them are made up only of what looks like faint haze. The really brilliant comets which have been seen now and then all through history (not, unfortunately, in the recent past) have very eccentric orbits, so that they take hundreds, thousands or perhaps millions of

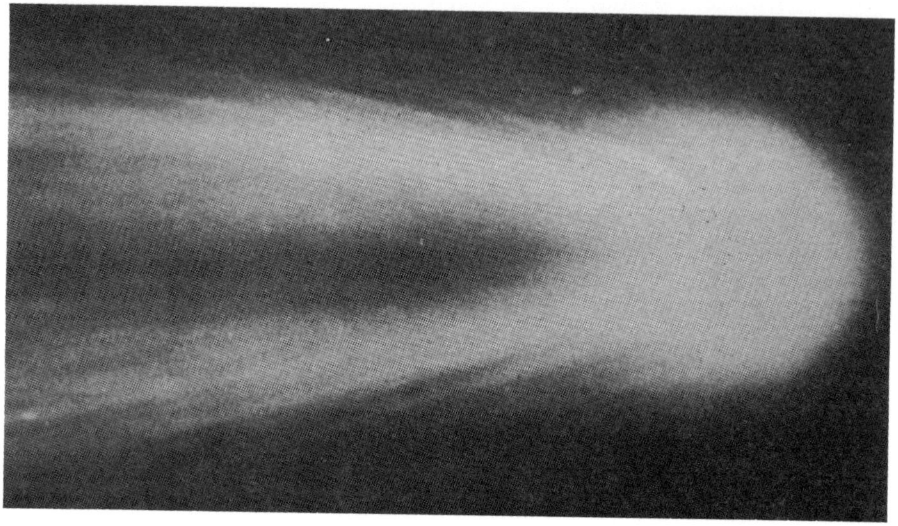

years to complete one circuit of the Sun and we cannot predict them. Tail or tails can develop only when the comet is relatively close to the Sun, and the tail always points away from the Sun because the dust and gas particles which make it up are driven outward by the solar wind. When a comet is receding from the Sun it is travelling tail-first. Halley's Comet will be moving in this way during the probe rendezvous, because perihelion will have been passed on the previous February 9.

If a comet loses a certain amount of material, by evaporation, each time it passes through perihelion, it seems that comets must have comparatively short lives. This does not apply to the comets which make only one revolution in an immensely long period, but there are many comets which return frequently; thus Encke's Comet has a period of only $3\frac{1}{3}$ years and has now been seen at more than 50 returns (actually, it can be followed throughout its orbit by means of modern equipment, because its distance from the Sun is never as great as that of Jupiter). We known of several periodical comets which have died. One was Biela's Comet, which was seen at intervals of $6\frac{3}{4}$ years until 1845, when it split in half. The two fragments were seen for the last time in 1852, after which they vanished as effectively as the hunter of the Snark, though meteors were subsequently observed coming from the position where Biela's Comet ought to have been. Another periodical comet which has certainly faded out is Brorsen's; and Westphal's Comet, which had a period of 61 years and was once visible with the naked eye, has also joined the ranks of the dear departed.

But what happens to a comet which has come to the end of its career? The masses are very low by planetary standards, and even a comet with a head or coma larger than the Sun (as with the Great Comet of 1843) is too lightweight to have the slightest effect upon the Earth or any other planet. It may well be that the debris visible in the form of meteors is all that is left once all the volatiles have been evaporated, though, as we have seen, it has also been suggested that the Apollo-type asteroids may be ex-comets.

IRAS, the Infra-Red Astronomical Satellite, detected asteroid Phæthon whose perihelion lies within the orbit of Mercury, and which moves in almost the same path as the Geminid meteors. Two more were also found; 1983 SA, which has a period of 8.6 years and an orbit which carries it from that of Mars out beyond the main asteroid belt, and 1983 XF, which comes nearly as close-in as the Earth but recedes to between the orbits of Jupiter and Saturn. An asteroid discovered much earlier, Hidalgo (No 944) travels from the distance of Venus out to nearly as far as Saturn. Telescopically it looks absolutely sharp, with no trace of 'fuzziness', but there are also some comets, such as Arend-Rigaux and Neujmin II, which look stellar now even though they formerly showed cometary characteristics. And studies of Halley's Comet, carried out at the Kitt Peak Observatory in Arizona during the early part of 1984, indicate that the comet is 'redder' than expected, so that it bears some resemblance in its make-up

Above left *Halley's comet in 1910.*

Right *Giotto, the Halley's comet probe, as I photographed it at British Aerospace when it was almost ready to be sent away for completion and testing.*

to the Trojan asteroids, which move in the same orbit as Jupiter, and to Phœbe, the outermost satellite of Saturn, which moves in a retrograde or wrong-way direction and may well be a captured asteroid rather than a bona-fide satellite.

All this is very interesting, and we may hope to learn more when the various space-craft encounter Halley's Comet in March 1986. Though the Americans cancelled their Halley probe on the grounds of expense, they do plan less important missions. They have altered the path of an old Earth orbiting satellite, originally known as ISEE 3, which was launched in August 1978; it is now to be sent to make a fairly close pass of a periodical comet known as Giacobini-Zinner. The manœuvre was a complicated one, involving several close approaches to the Moon, but it seems to be successful. NASA also means to send a probe to Kopff's periodical comet in 1994. Kopff's Comet (never bright enough to be visible with the naked eye) is believed to be 'dusty' and probably not very ancient by cosmical standards. The space-craft will contact the comet when it is still in the asteroid belt, and will come within ten miles of it. Then, as Kopff's Comet approaches the Sun, it will start to become active; the space-craft will 'back off' to some distance and continue operating, matching the comet's speed round the Sun and staying with it for several years. The procedure is quite different from that with Halley's Comet because Kopff moves round the Sun in the same sense as the Earth, while Halley is retrograde, so that the probes will have to meet it practically head-on.

As a bonus, the Kopff mission will also take in two small asteroids, Namaqua (about 8 miles in diameter) and Lucia (28 miles). Whether this will be the first asteriod rendezvous remains to be seen.

So, all in all, the coming years will be of special interest. As a spectacle, Halley's Comet may be rather a disappointment, because it will not be as brilliant as it was in 1910, but it has much to tell us. There is no doubt that it will become the best-observed comet of all time!

48 THE ORANGE WORLD OF TITAN

On November 12 1980 the space-craft Voyager 1 passed by the magnificent planet Saturn at a distance of less than 80,000 miles, sending back the first really detailed pictures of the surface and the rings. But Saturn was not Voyager's only target. It had already by-passed Jupiter, and it was also concerned with Saturn's system of satellites. In particular, it was programmed to study Titan.

Jupiter has four large satellites (the Galileans) and a dozen very small ones. Saturn, on the other hand, has one really major attendant, Titan, and several of medium size; Rhea and Iapetus are between 900 and 1,000 miles in diameter, Dione and Tethys between 600 and 700. The remaining members of Saturn's family are much smaller. Titan, with its diameter of 3,200 miles, is larger than the planet Mercury, and of known satellites only Ganymede in Jupiter's system is larger. Telescopically, Titan appears as a tiny disk but even before the flight of Voyager 1 it was known to be an exceptional world. Alone among Solar System satellites, there was proof of the existence of a substantial atmosphere.

This atmosphere was first reported as long ago as 1908 by the Spanish astronomer Comas Solá, who saw that Titan showed the phenomenon of 'limb darkening'; that is to say, the edges of the disk appeared less bright than the centre, because the light from the solid globe had to pass through a thicker layer of Titan's atmosphere. Then, in 1944, G. P. Kuiper had detected the atmosphere spectroscopically. It was tacitly assumed to be much thinner than that of the Earth, and to be made up chiefly of the gas methane (more familiarly known as marshgas). Titan's escape velocity is only 1½ miles per second, which is much the same as that of our airless Moon; but Titan is much further from the Sun, and the lower temperature means that the atoms and molecules in its atmosphere move more slowly, so that they are easier to hold.

Subsequently, astronomers began to have second thoughts. There was some evidence that Titan was cloud-covered, in which case it would be a very interesting place — so interesting, indeed, that it led to a modification in the Voyager missions. Voyager 1 was due to make a close pass of Titan. If it succeeded, it would simply go on out of the Solar System, to begin a never-ending journey between the stars. If it failed, then its follower, Voyager 2, would undertake the Titan survey — in which case it would be unable to go on to rendezvous with the outer giant planets, Uranus and Neptune. Therefore there were fervant hopes that Voyager 1 would succeed. I was at the headquarters of the mission, the Jet Propulsion Laboratory at Pasadena in California, and as Voyager 1 drew in toward Saturn and its system, the excitement mounted. The rings provided shocks aplenty; there were thousands of ringlets and narrow gaps. But would we be able to see the actual surface of Titan? Opinions differed, but before long the answer became obvious; we would not. Voyager passed by Titan at a distance of only 4,000 miles, but nothing could be seen apart from an orange disk, rather darker in the north than in the south. Like Venus, though for quite different reasons, Titan hid itself with devastating completeness. One lovely picture showed a crescent Titan, with the cusps prolonged as the upper atmosphere caught the sunlight — a phenomenon also familiar to any regular observer of Venus.

Efforts were made to detect a magnetic field. They failed, and evidently Titan's own field is negligible but, perhaps significantly, it moves round Saturn at a distance of 760,000 miles, which is just about the limit of Saturn's main magnetosphere. Titan lies sometimes inside the magnetosphere, sometimes outside it.

The relative darkness of the northern hemisphere was interesting. Titan has a captured or synchronous rotation; it takes 15.5 days to spin on its axis, and this is the same as its revolution period, so that Titan keeps the same face turned toward Saturn all the time. (This is also true of the other satellites, apart from Phœbe.) However, it is more likely that the darker hue of the northern

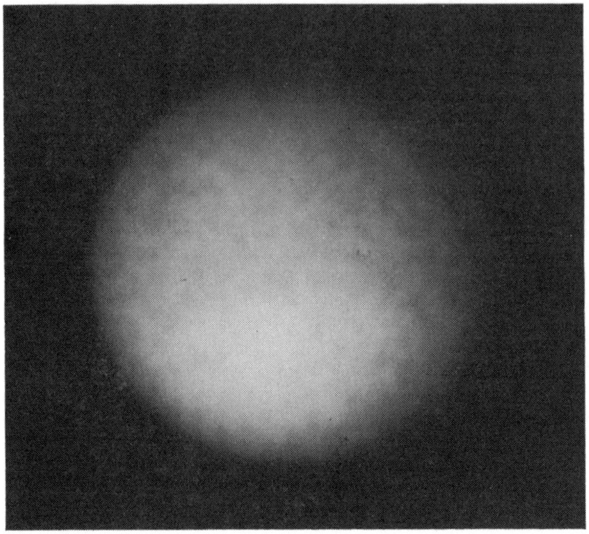

Titan — photographed from Voyager 1.

hemisphere is some kind of seasonal effect; remember, Saturn takes over 29 years to complete one journey round the Sun, and at present it is the planet's northern hemisphere which is tilted sunward.

False-colour processing brought out the extensive haze above the main cloud-deck, but it was the composition and the density of Titan's atmosphere which provided the main surprises. First, the ground pressure is 1.6 times that of the Earth's air at sea-level, so that the atmospheric density is considerable — compare it with that on Mars, where the pressure is so low (around 7 millibars) that it corresponds to what we would normally call a vacuum. And instead of being chiefly methane, the atmosphere of Titan turned out to be mainly nitrogen, which of course makes up 78 per cent of our own air. On Titan there is about 6 per cent methane, together with other compounds such as ethane and acetylene, plus a certain amount of helium; but nitrogen is dominant.

We can now make a good estimate of the structure of Titan's atmosphere. Above the surface there is probably a clear region, and then come the methane clouds. Above this there is a layer in which Voyager 1 detected aerosols ('smog', if you like) made up of various ammonia/methane compounds. Overlying the cloud-deck, which we can see, there is the upper haze, and still higher, at 250 miles, a layer which absorbs ultra-violet light.

What about the globe itself? The overall density is just under twice that of water, so that there must be ice mixed in with the rocky material; probably rock accounts for about 55 per cent. Presumably there is a relatively heavy core, above which comes a deep, icy layer which is overlaid by the rock and ice crust. About the surface itself we have no direct information, but we can at least speculate.

According to Voyager, the surface temperature is about -168°C. Now this is near the triple point of methane — that is to say the temperature at which methane can exist as a solid, a liquid or a gas, just as H_2O can exist on Earth as water, ice or water vapour. Methane melts at -182°C and boils at -155°C, so that the surface temperature of Titan falls neatly in between these two limits. There may be oceans of liquid methane, cliffs of solid methane, and a steady drip-drip of methane rain from the orange clouds. The winds in the upper atmosphere may be reasonably strong, perhaps up to 200 knots, but they could well be less violent near the surface, as is the case on Venus.

If an ocean exists, it may be made up largely of substances such as ethane, but there is also an intriguing suggestion due to Carl Sagan and Stanley Dermott, in America. They believe that there may be a vast methane ocean at least 1,000 ft deep! According to this theory, the methane gas in Titan's atmosphere comes from a source on the surface, which is presumably liquid. Each time Titan orbits Saturn there should be a 30 ft tide. This would cause friction between the ocean and the ocean bed, and in time Titan would be forced into an orbit where the friction has been reduced to zero. Yet this has not happened, so that the friction must be very slight. In a shallow ocean, currents are faster and friction greater than in a deep one. So, say Sagan and Dermott, either Titan has no ocean at all, in which case the atmospheric methane is hard to explain, or else the ocean is very deep. Titan could have a reservoir of natural gas 200 times as great as that of the Earth.

It is a fascinating idea, and it makes one wonder whether future astronauts will need something in the nature of a cosmic submarine rather than an ordinary capsule! Certainly Titan is utterly unlike any other world in the Solar System. Organic compounds unquestionably exist, and there are in fact all the ingredients for life, but probably the very low temperature has prevented life from appearing there; Titan has even been described as an Earthlike world in deep freeze.

Eventually things will change. When the Sun turns into a Red Giant star, as it will do in perhaps 5,000 million years from now, life on Earth will be destroyed, and the outer worlds will be warmed up. Unfortunately this does not mean that Titan is likely to become habitable. First, the Sun's Red Giant stage will not last for long enough for life to evolve. Secondly, once Titan is warmed it will lose its atmosphere.

Voyager 2 did not make a close approach to Titan; it could have added very little, and so it could afford to leave Titan alone and move outward toward Uranus (1986) and Neptune (1989). As no more Saturn probes are planned as yet, we may have a long wait before we learn more. It is just possible that Earth-based radar may be able to decide whether or not Titan's surface is liquid, when Saturn nears perihelion in the 1990s, but otherwise we will have to wait for a Titan probe, put into a path round the satellite to map the surface by radar as the latest probes have done with Venus.

In any case, Titan is unique. When you look at it through a telescope, and see it as a tiny dot of light, it is not easy to visualize what may be the scene in the bitterly cold ocean of Titan under its gloomy orange sky.

49 GREENWICH TIME

October 13 1984 was an important anniversary. Exactly 100 years ago Greenwich Mean Time was adopted on an international basis and the Greenwich meridian was accepted as marking the zero for longitude, so that it divides the world into the eastern and the western hemispheres.

Even though international agreement was a great deal easier to obtain a century ago than it is now, there were many arguments and many heated discussions. Everything hinged upon the importance of accurate timekeeping, and it was the reputation of Greenwich in this respect which led to the final decision.

The original clocks were, of course, very simple; clepsydræ or water-clocks, candles burning at a known rate, and so on. Mechanical clocks came much later. The clock in Salisbury Cathedral was made in or before 1386, and is almost certainly the oldest mechanical clock in the world still in working order; it has no dial or hands, but it does ring a bell to give the time. The pendulum clock was developed by Christiaan Huygens in the 1660s. Today we have atomic clocks which are better timekeepers than the Earth itself. But so far as Britain was concerned, the main problem many years ago was to find a method by which sailors could find their position when out of sight of land, for this accurate clocks were needed.

The clock in Salisbury Cathedral.

Above left *The 'Onion Dome' of the Greenwich 20 in refractor with Patrick Moore.* **Above right** *The dome of the Greenwich refractor—open on Longitude Day.*

Finding one's latitude is easy enough. All that has to be done is to measure the altitude of the Pole Star, and make an adjustment to allow for the fact that the star is not exactly at the pole — the height of the pole above the horizon is equal to the observer's latitude. Longitude was more troublesome. Astronomical observations can provide the local time, and comparing this with Greenwich time tells how far the observer is from the Greenwich meridian — in other words, his longitude. But this means knowing Greenwich time, and earlier clocks were not capable of keeping correct time at sea. Huygens went on a voyage to test out two pendulum clocks, but all that happened was that the bob of one pendulum fell off, the other clock stubbornly refused to work and Huygens' assistant was violently seasick!

Longitude-finding continued to be troublesome, and led to several disasters, including one in 1707, when four ships of Queen Anne's Navy were wrecked on the Scilly Isles because the Admiral in charge lost his way and turned in the wrong direction. The problem was finally solved by the invention of the chronometer by John Harrison later in the 18th century. Harrison's chronometers are still in working order; you can see them today in the National Maritime Museum.

Left *The time-ball at Greenwich.*

Right *Herstmonceux Castle — home of the modern Royal Greenwich Observatory.*

However, it was only with the coming of the railways that the need for standardizing time-systems became apparent. Even over a country as small as England, there is a definite difference in local time between, say, London and Bristol; in the days of stage-coaches it was customary for the driver to use an adjustable watch, running fast on the eastbound London journey and slow on the westbound Bristol journey, but from 1825, when the first passenger train service came into operation, things were less easy. There was a classic case of confusion in July 1858, during a court case in the town of Dorchester. The case was timed to start at 10 am, but the defendant was not present and in his absence the verdict went against him. He duly turned up some minutes later — because he had been reckoning by one time-system and the court by another. On appeal, it was decreed that 'ten o'clock must be considered ten o'clock according to the time of the place. The difference between Greenwich time and Dorchester time would be about 9 minutes. If the clock were set by Greenwich time it was not 10 o'clock when the case was tried, and the defendant ought to be let in to try it.' I have never been able to find out the final result of the case!

Obviously it would be convenient to adopt 'railway time' everywhere, and by 1840 the Great Western Railway was already doing so. The city of Oxford went so far as to fit two minute hands on to its Great Tom Clock in Christ Church — one giving Greenwich time and the other Oxford time. Finally, in 1880, 'railway time' was re-named Greenwich Mean Time, and made universal for the whole of Britain. In 1884 it was also adopted by most nations. (France, not surprisingly, objected, and adopted 'Paris time', which was 9 minutes 21

seconds different from Greenwich; in 1911 a change in French law decreed that legal time there would be Paris Mean Time retarded by 9 minutes 21 seconds — in other words Greenwich Mean Time, which was exactly the same thing but which satisfied French national pride.)

The problems in America were more acute, because the coast to coast difference in local time over the United States is as much as three and a half hours. The solution adopted was due to Charles Dowd. The USA was divided into definite time zones, based on Greenwich time; all the minute hands would remain the same, and only the hour hands would alter from one zone to another.

On October 1 1884 the International Meridian Conference was convened in Washington, 'to discuss and if possible fix upon a meridian proper to be employed as a common zero for longitude and standard of time reckoning throughout the whole world'. Greenwich was the obvious choice. It was here that Sir George Airy, the somewhat forbidding Astronomer Royal who had held office between 1835 and 1881, had set up his transit instrument, and it was this instrument which was proposed to mark longitude 0°. The only nation voting against the proposal was San Domingo, though France (of course!) and Brazil abstained.

Since then there have been no objections to the use of Greenwich time and the Greenwich meridian, even though the Royal Observatory has been moved from its old site to the clearer skies of Herstmonceux in Sussex, and the old observatory has been turned into a museum. The Airy transit instrument is no longer used (though it is still fully functional, as I found out when I made some test observations with it a few years ago) and the modern atomic clocks at Herstmonceux are almost incredibly accurate. We still reckon from Greenwich itself rather than Herstmonceux, which is not exactly on the zero meridian, but the actual signals are generated at Herstmonceux, which is accepted as the headquarters of 'world time'.

It was navigation which called for the development of really accurate clocks, and fast travel which necessitated the standardisation of time systems. In Britain, the Victorian trains and postal services, unlike those of today, really kept to their schedules, and there is one amusing demonstration of how things have changed. In this centenary year of the Greenwich meridian, a repeat of an old stage-coach mail run from Bristol to London was organised, and letters were duly brought from the one city to the other. As the stage-coach left Bristol, an unkind observer posted a letter to London in a nearby box, taking care to use a first-class stamp. The letters carried on the stage-coach arrived in London first.

50 REFLECTION NEBULÆ

Have you ever watched a car coming up a misty road, on a dark night, with its headlights full on? If so, you will have noticed that around the lights there is a definite halo. This is due simply to the particles of mist being illuminated. We see the same sort of appearance in the sky, making up what are termed reflection nebulæ.

Nebulæ are of two main types. First there are the emission nebulæ, of which the best-known example is in the Sword of Orion; it is easily visible with the naked eye, below the three bright stars of the Hunter's Belt, and it is known to be a stellar birthplace, where fresh stars are condensing out of the nebular material. The gas is made up largely of hydrogen, which is the most plentiful substance in the entire universe. The hydrogen atoms are split up or ionized by radiation from the very hot stars embedded in the nebula, and when these atoms recombine they give off light. Emission nebulæ are common enough; some 5,000 million years ago our Sun was presumably born inside one of them.

True reflection nebulæ are quite different. They are made up largely of 'dust', and the material is not necessarily associated with the star making them visible; the dust-grains merely happen to be suitably placed. The grains are extremely small (on average, about a millionth of a metre in diameter) and they are by no means packed closely together; the usual density amounts to the equivalent of one grain in an area the size of the interior of St Paul's Cathedral. All the same, even matter as thinly-spread as this is able to make itself evident.

Viewed with the eye at the telescope, reflection nebulæ look white, but when photographed they show lovely colours — which are not visible directly simply because the level of illumination is too low, just as it is impossible to distinguish colours of everyday objects by moonlight. Superb pictures have been taken by David Malin with the great reflector at the Anglo-Australian Observatory at

Above *The Lagoon Nebula, Messier 8, photographed in red light with the Palomar 200 in reflector.*

Below *The Cone Nebula in Monoceus. Photographed in red light with the Palomar 200 in reflector.*

The nebula Messier 16 in Serpens.

Siding Spring in New South Wales, and his new techniques have proved to be remarkably effective. Some of the reflection nebulæ show up as blue, and to explain why this is so it will be helpful to consider the colour of our own sky during the daytime. The atmosphere 'spreads around' the blue part of the Sun's light, which is more easily scattered than the longer-wavelength red part. This not only makes the sky blue, but also makes the Sun look slightly redder than it would otherwise do. In fact, it is fair to say that the Sun is 'de-blued'. Exactly the same principle applies to reflection nebulæ, and the illuminating stars may look reddish even when they are not genuinely so.

Sometimes emission and reflection nebulæ appear side by side. A good example of this is the famous Trifid Nebula in the constellation of Sagittarius, the Archer, known officially as M 20 because it was the 20th object in a famous list of star-clusters and nebulæ drawn up by the French astronomer Charles Messier more than 200 years ago. M 20 itself is a typical emission nebula, shining because of the hot stars inside it; it is distinguished by the prominent dark lanes which have given it its familiar nickname. David Malin's photographs show a blue reflection nebula adjoining it. Apparently the Trifid itself is embedded in a dusty sphere, and it is this which produces the blueness.

If the dusty material of a reflection nebula is not directly associated with the illuminating star, we have to assume that the star simply moves around and sometimes enters a dusty region, lighting it up until it emerges once more — a process which will, of course, take a very long time indeed. The light from the star is not directly blocked out, but is deflected by the process which is termed forward scatter — just as happens with the car headlights on a dark, misty road. Obviously, stars of all kinds can produce reflection nebulæ, and some of these stars are truly red. A classic case is that of Antares, in the Scorpion, which is a highly luminous Red Supergiant big enough to contain the whole orbit of the Earth round the Sun. It lies in a very dusty region, and the Australian photographs show gloriously coloured nebulæ in the whole area.

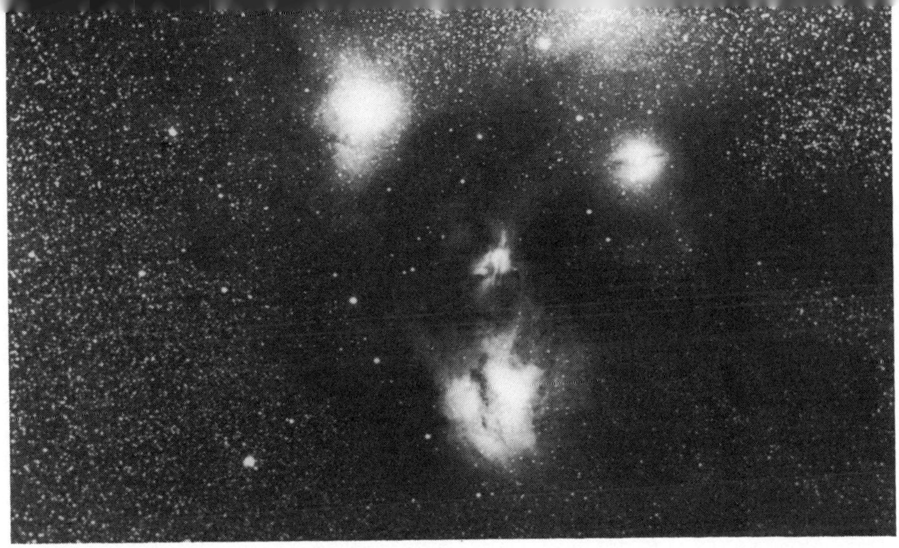

Nebulosity round the star Rho Ophiuchi.

On the other hand there are some exceptional objects which come into a different category. One of these is the so-called Toby Jug Nebula; the nickname comes from its somewhat jug-shaped appearance. Here we have a very old, red star which is coming to the end of its main career, and is puffing out material in all directions, so that for once the reflection nebula really is being manufactured by the star inside it. It has been carefully studied at infra-red wavelengths, and the results are somewhat surprising. The material seems to be made up of silicates; in other words, sand, so that the old star has surrounded itself with a sandy mantle. Most of the thinly-spread interstellar 'dust' seems to be in the form of graphite, so that the Toby Jug is very unusual indeed, though some similar nebulæ are known. Unfortunately it is a faint object, and its precise nature was uncertain until the recent work carried out from Siding Spring.

Reflection nebulæ are very important to modern astronomers. By now we have a fairly good idea of the life-stories of average stars, but we are less certain of the earliest and latest stages. Stars of moderate mass, such as our Sun, expand when they use up their reserves of nuclear 'fuel' and throw off their outer layers, producing what are termed planetary nebulæ — a bad name, since a planetary nebula is not truly a nebula and is certainly not a planet! After this stage, the remnant of the star collapses into a very small, extremely dense White Dwarf. This will happen to the Sun eventually, though not for several thousands of millions of years yet. A more massive star may explode in a tremendous outburst known as a supernova, with its remnant forming an even smaller, denser object made up of neutrons; the classic case of this is the Crab Nebula in Taurus, the Bull. With an even more massive star, the end product may be a Black Hole, pulling so strongly that not even light can escape from it.

Studies of reflection nebulæ may give us vital clues about processes of this kind. And quite apart from this, the new techniques provide magnificently beautiful views of them. With their glorious hues and their intricate studies, they are as lovely as any objects in the whole of the sky.

INDEX